Virtual Teamwork

VIRTUAL TEAMWORK

Mastering the Art and Practice of Online Learning and Corporate Collaboration

EDITED BY ROBERT UBELL

A JOHN WILEY & SONS, INC., PUBLICATION

Published by John Wiley & Sons, Inc., Hoboken, New Jersey
Published simultaneously in Canada

For general information on our other products and services or for technical support, please contact our Customer Care Department within the United States at (800) 762-2974, outside the United States at (317) 572-3993 or fax (317) 572-4002.

Wiley also publishes its books in a variety of electronic formats. Some content that appears in print may not be available in electronic formats. For more information about Wiley products, visit our web site at www.wiley.com.

Library of Congress Cataloging-in-Publication Data:

Virtual teamwork : mastering the art and practice of online learning and corporate collaboration / edited by Robert Ubell.
p. cm.
Includes index.
ISBN 978-0-470-44994-3 (pbk.)
1. Virtual work teams. 2. Teams in the workplace. 3. Organizational learning. I. Ubell, Robert.
HD66.V558 2010
658.4'022–dc22

2009049264

Printed in the United States of America

10 9 8 7 6 5 4 3 2 1

For Rosalyn

Contents

ACKNOWLEDGMENTS

This book is partly the result of more than a decade of engagement with online learning that began serendipitously on a visit to Stanford University, where I came upon a remarkable activity that very few had ever witnessed before. It was there, at the Stanford Center for Professional Development in the School of Engineering, where I met Andy DiPaolo and his pioneering colleagues who were launching something, I sensed, that was not merely an experiment suitable for elite students only. I was observing the start of what was then being tentatively practiced in only a handful of universities where online learning was being born. Today, it is part of university life, with some 4.6 million US students online.

I flew back to New York, excited by what I had seen, eager to tell my colleagues all about it. At the time, I was working for a small scientific and technical publisher as head of new media. Thinking the company would grasp the implications of what Andy and his colleagues were doing, I proposed that the president consider entering into a partnership with Stanford—we would publish ancillary print materials, while the university would deliver its courses online.

"Online learning has no future," the company's president predicted.

It didn't take deep reflection to recognize that my enthusiasm was unlikely to be supported, and so I retreated. Luckily, soon afterward, I stumbled on an advertisement in *The Times* seeking someone to head a new venture in "web-based distance learning" at Stevens Institute of Technology, an engineering school just across the Hudson in New Jersey. With no experience—except for my few hours at Stanford—I applied. It turned out that it was all that was required. A dozen years ago, hardly anyone knew anything about online education, so my meager exposure and my enthusiasm were enough of a résumé.

It wasn't long before I attended a symposium in midtown Manhattan, sponsored by the Alfred P. Sloan Foundation, where I met Frank Mayadas, program director for what the foundation called "asynchronous learning networks," an unlikely name for what was later called "e-learning."

"Join me for lunch," Mayadas encouraged. We set a date to meet at an Indian restaurant not far from Rockefeller Center, where the Sloan Foundation has its headquarters. Over curry and dal, I sketched what I was doing—hoping to create an online graduate program in science, engineering, and management. But it was early days and I was struggling with tepid faculty response at best, hostility at worst. Only the most adventurous had agreed to participate, with merely 3 courses and 23 students enrolled in the first semester.

While quite sympathetic, Frank nonetheless proposed that I move forward even more boldly. By the time we took our last bites of watermelon for dessert, Frank offered me a Sloan grant to launch an online master's degree in wireless communications, a graduate program that was being offered at Stevens on campus but had not yet migrated online.

With the promise of Sloan Foundation funding in my pocket, I took a PATH commuter train back to Stevens' campus in Hoboken. Elated, I nearly flew into the office of the head of the graduate school with the news. "You must have misunderstood," he shook his head. "Nobody gets a grant over lunch."

Frank's unexpected generous investment was among the early encouragements that eventually led WebCampus, Stevens' online graduate program, to attract nearly 25,000 enrollments over 10 years. This book would never have been possible without Frank Mayadas' foresight and personal and professional support and friendship. Without Frank's evangelism and the Sloan Foundation's seed money, online learning would not now be practiced by some 4.6 million college and university students who take at least one online class. Online learning students represent nearly 25% of all students in higher education.

I am grateful to my many colleagues whom I had encountered owing to my engagement with the Sloan Foundation. For their generosity in sharing their knowledge and insight, I owe my appreciation to Eric Fredericksen, University of Rochester; Jacquie Maloney, University of Massachusetts; Tony Picciano, City University of New York; Peter Shea, University of Albany; and Karen Swan, University of Illinois.

Members of the Board of Directors of WebCampus tirelessly offered their wisdom and experience, helping the program achieve national and international recognition. For their invaluable support as well as for their personal commitment, I thank Phil Long, University of Queensland; Luther Tai, Consolidated Edison; Peter Wiesner, IEEE; Kee Meng Yeo, Amway; and Ronald Schlosser and Martin Tuchman. I am also enormously grateful to Stevens' faculty and staff who served on the board as well as those who were especially supportive as WebCampus professors—Larry Bernstein, Stephen Bloom, Hong-Liang Cui, Celia Desmond, Peter Dominick, Sven Esche, Hosein Fallah, Thomas Herrington, John Horgan, Peter Jurkat, Dilhan Kalyon, Donald Lombardi, Manu Malek, Barbara Migliori, Ann Mooney, Barbara O'Connor, Richard Reilly,

Kevin Ryan, Steven Savitz, Charles Suffel, B. J. Taylor, and Yu-Dong Yao.

At the Polytechnic Institute of New York University, President Jerry Hultin has been enormously supportive, easing the way for me and my staff to build on the e-learning foundations established earlier at the school. To Jerry and his colleagues at the Polytechnic Institute, I owe an enormous debt for their welcome and encouragement. I offer my deep appreciation to Lilana Avery, Kurt Becker, John Bernhard, Ria Best, JeanCarlo Bonilla, Lea Bowie, George Bugliarello, Brian Casey, Crystal Chavis, Ji Mi Choi, Jonathan Chao, Joy Colelli, Mary Cowman, Dennis Dintino, Dawn Duncan, Alan Fisher, Robert Flynn, Andres Fortino, Barbara Kates Garnick, Steven Goss, Ardis Kadiu, Iraj Kalkhoran, Meera Kumar, Sunil Kumar, Erich Kunhardt, Joseph Lathan, Marlene Leekang, Kalle Levon, I-Tai Lu, Nasir Memon, Peter Morales, Felice Nudelman, Shivendra Panwar, Susan Puglia, Bharat Rao, Dianne Rekow, Jana Richman, Keith Ross, Carl Skelton, Kate Smith, Harvey Stein, Richard Thorsen, Jay VanDerwerken, Philip Venables, Yao Wang, Nina Weber, Allan Weisberg and T. C. Westcott.

For her skill and resourcefulness, I owe my warmest thanks to my steadfast colleague and friend, Elaine Cacciarelli, whose dependable, consistent, and careful attention to details made this book possible.

One of the most pleasant experiences during the making of this book has been my happy relationship with the skilled staff at John Wiley & Sons. I became warmly reacquainted with Anita Lekhwani who, with her expert staff, shepherded this book to publication. I owe her a great debt of gratitude for her enthusiastic agreement to go ahead with this project and for her eager encouragement throughout. I especially wish to thank Rebekah Amos, Kellsee Chu, Kim McDonnell, Sheik Safdar, and Sanchari Sil at Wiley who helped to bring this book to publication.

I am particularly indebted to the authors of chapters in this book who explored their rare experiences in virtual classrooms with sharply honed intelligence. Faculty and students who study their insights will come away with a deep appreciation for what it takes to participate in virtual teamwork. Their brief professional biographies appear elsewhere in the opening pages.

I am deeply indebted to my family for their love and affection—to Jennifer, Thornton, Ella, Ben, Elizabeth, Steve, Fordon, Marielle, Matt, Jack, Shane, and Bryn, Seymour, Marsha, Anne, Marvin, Alvin, Estelle, Evelyn, Stella, and Ernesto; and to my friends—Robert Benton, Hal Espo, Martha Gever, Andrea Marquez, Robert Millner, Yvonne Rainer, Florence Rowe, Neil Salzman, Sheila Slater, and Stephen Stanczyk. I owe so much to my brother Earl that it is impossible to express the loss I feel without him.

This book is dedicated with my deepest love to an extraordinary woman, who is not only a committed scholar, stepmother, grandmother, and friend but is also someone who turns ordinary days into an examined and purposeful life.

—Robert Ubell
New York, 2010

CONTRIBUTORS

Phylise Banner

Phylise Banner is an instructional design project leader at the American Public University System.Banner works with faculty, administrators, students, and IT managers to design, develop, and deliver technology-based solutions across departments and disciplines. Using emerging technologies to translate faculty pedagogical approaches into unique online learning environments outside of traditional course management systems, she currently focuses on the use of Web 2.0 applications to establish community and visual presence in online classrooms.

M. Katherine (Kit) Brown-Hoekstra

Kit Brown-Hoekstra is the principal of Comgenesis, LLC, providing consulting services and training to clients to produce internationalized documentation, as well as other consulting services. She is an award-winning writer with a background in life sciences and technical communication. Brown-Hoekstra contributes articles and speaks at conferences and workshops worldwide on a variety of technical communication topics. She is an Associate Fellow of the Society for Technical Communication, newsletter editor for the IEEE-PCS, and member of the American Medical Writers Association.

Richard Dool

Educator, consultant, and senior executive, Richard Dool has diverse senior management experience including as a CEO of both public and private companies. He has a range of management experiences including leading an $800 million division of a Fortune 20 company, rescuing a company from near bankruptcy, leading the acquisition or divestiture of 9 companies, and managing companies in the UK, Germany, France, Spain, Hong Kong, India, and Australia. He is on the faculty at Seton Hall University and teaches and directs the graduate communication program. His research and publication interests are in leadership communication, distance learning, and change management. His publications include *Enervative Change: The Impact of Persistent Change Initiatives on Employee Job Satisfaction* (2007). He is also a Sloan-C Certified Online Instructor. He is the editor-in-chief of *Influere: The Leadership Communication Journal*. Dool received his DMgt from the University of Maryland University College.

Brenda Huettner

Brenda Huettner is owner of P-N Designs, Inc., a communication consulting company. She writes articles and teaches workshops on management, usability, and technical writing. Huettner is a Fellow of the Society for Technical Communication and belongs to the Southern Arizona Chapter STC and the Management, Independent Consulting and Contracting, Usability, and AccessAbility SIGs.

Char James-Tanny

Char James-Tanny is president of JTF Associates Inc. A technical writer well known in the Help community for her knowledge of online Help tools and concepts, she speaks frequently at conferences around the world on Help topics, cross-browser issues, and tool-specific functionality.

James-Tanny is an Author-it Certified Consultant and a 2010 Microsoft Help MVP. She is also the secretary of the Society for Technical Communication.

Elaine Lehecka Pratt

Elaine Lehecka Pratt is president of Lehecka Pratt Associates, Inc., a regulatory compliance consultancy providing training to the pharmaceutical, medical device, and biotech industries. She currently serves on the PDA planning committee for the biennial training conference, and has previously served as president of the GMP Training and Education Association, cochair of the American Society for Training and Development Pharmaceutical-Chemical Industry Group, and on the PMA Training and Education Resource Committee. She currently serves on the PDA planning committee for the biennial training conference, and has previously published in *Pharmaceutical Technology, Pharm Tech Japan, Performance in Practice,* and *Drug Development and Industrial Pharmacy,* and is a frequent speaker at national and international meetings. She is on the faculty of the PDA Training and Research Institute, and is an industry professor at Stevens Institute of Technology, where she teaches in the graduate programs of pharmaceutical manufacturing and pharmaceutical management. Lehecka Pratt holds a BS in biology from Ursinus College and an MBA in management from Farleigh Dickinson University.

Haekyung Lee

Haekyung Lee is a scholar with expertise in instructional system design, web-based learning environments, e-learning, computer-supported collaborative learning, collaborative learning, computer-mediated communication, and peer and self-assessment. She received her BS and MS degree in educational technology from Ewha Woman's University in Korea. She cofounded ComLiving.com Corporation, a provider of web-based online education and developer of online content. Lee has worked as an instructional designer, a project manager, and a consultant for designing and developing online training courses. Since she earned her PhD from The University of Texas at Austin in 2008, she has been working as a human resource development specialist at CJ Corporation in South Korea.

Paul E. Resta

Paul E. Resta holds the Ruth Knight Milliken Centennial Professorship in Instructional Technology and serves as director of the Learning Technology Center at The University of Texas at Austin, where he teaches advanced graduate courses in instructional technology. His scholarly work covers educator development and technology, digital equity, and the design of web-based learning environments. Resta serves as chair of the International Jury for the UNESCO Prize for ICT in Education and the Association of Teacher Educators' National Commission on Technology and the Future of Teacher Education. He is also founding president of the International Society for Technology in Education and served as president of the International Council of Computers in Education. He has held positions with the US Department of Education, University of New Mexico, and the Systems Development Corporation. Resta has received top honors from the Society for Information Technology in Teacher Education, the US Distance Learning Association, and the University Continuing Education Association, among many professional groups. Resta received his PhD from Arizona State University.

Anne-Françoise Rutkowski

Anne-Françoise Rutkowski is associate professor of information systems and management at Tilburg University in the Netherlands. She was awarded her PhD in social and cognitive psychology from the same university in the Department of

Psychology. Her research interests and publications bridge IS and human sciences in addressing topics such as group decision making, processes of attention with ICTs, overload, e-learning, virtual and multicultural collaboration in teams. Her work has been published in such journals as *IEEE Computer*, *Decision Support Systems*, *Group Decision and Negotiation*, and *Small Group Research*.

Michael Ryan

Michael R. Ryan, a clinical professor and executive director of the Institute for Leadership Research at Texas Tech University, received his PhD in technology management with a concentration in organizational behavior from Stevens Institute of Technology. His research interests cover leadership development, team development, product innovation, and project management, with recent work on leadership in virtual teams. Before entering his academic career, Ryan founded his own logistics company and started an information technology venture in the Small/Medium Business market. As a consultant, he worked with numerous Fortune 500 companies, including AT&T, Lucent, and IBM. Prior to joining Texas Tech, Ryan was an adjunct faculty member at Stevens Institute of Technology, where he co-founded Stevens Institute for Technical Leadership.

Carol Stoak Saunders

Carol Stoak Saunders is a professor of management information systems at the University of Central Florida, where she performs research on the organizational impact of information technology, virtual teams, time, overload, sourcing and interorganizational linkages. She served as general conference chair of ICIS '99 and Telecommuting '96. She was also the chair of the Executive Committee of ICIS, inducted as an AIS Fellow and editor-in-chief of *MIS Quarterly*. She publishes in *MIS Quarterly*,

Information Systems Research, Journal of MIS, Communications of the ACM, Academy of Management Journal, Academy of Management Review, and *Organization Science.* Saunders received her PhD from the University of Houston.

Anu Sivunen

Anu Sivunen is research manager of the Virtual and Mobile Work Research Unit at Aalto University School of Science and Technology in Finland. Her research interests cover global virtual teams, collaborative virtual environments, group identity, and technology-mediated communication. Sivunen is working as a visiting scholar at Stanford University. Results of her research have been published in *IEEE Transactions on Professional Communication, Group Decision and Negotiation,* and *Journal of E-working.* Sivunen received her PhD in speech communication from the University of Jyväskylä in Finland.

Luther Tai

A lawyer and engineer, Luther Tai is senior vice president of Enterprise Shared Services at Consolidated Edison Company. He also oversees The Learning Center, Con Edison's corporate university. Earlier, Tai served as senior vice president of Central Services, senior vice president of Central Operations, and vice president of Corporate Planning. Tai serves on the board of Woodrow Wilson National Fellowship Foundation, on the board of Stevens Institute of Technology's WebCampus, and on the advisory board of the Executive Program in Work-based Learning Leadership at the University of Pennsylvania. He is also a member of the board of directors of the Regional Plan Association. Named a Fulbright scholar in 2006, he is the author of *Corporate E-Learning: Inside View of IBM's Solutions* (Oxford, 2007). He holds a BS in chemical engineering from MIT, a master's degree in industrial engineering from Columbia University, an MBA from Cornell

University, a JD from New York Law School, and a Doctor of Education from the University of Pennsylvania. He is also a graduate from Harvard Business School's Advanced Management Program.

Robert Ubell

Robert Ubell is vice president of Enterprise Learning at NYU Polytechnic Institute, where he heads the school's online learning unit, NYU ePoly. Earlier, he launched Stevens Institute of Technology's online graduate program, WebCampus, and administered the school's China program. Ubell was vice president and editor-in-chief of Plenum Publishing Corporation, editor of the National Magazine Award-winning monthly, *The Sciences*, and American publisher of the premier British science weekly, *Nature*. He was also founding publisher of *Nature Biotechnology*. Ubell was head of his own print and e-publishing consulting firm, Robert Ubell Associates, and is the author or editor of five books and more than 50 scholarly articles. Ubell serves as vice president of the board of the Parkinson's Unity Walk Foundation and is on the Sloan-C Annual Conference Steering Committee. He is chair of ASTD's New York eLearning Special Interest Group and is a member of the online learning board of Borough of Manhattan Community College. Ubell also serves on the board of the Sloan Consortium. He received his undergraduate degree from Brooklyn College and was a guest lecturer at MIT and Columbia University's College of Physicians and Surgeons.

Christine Uber Grosse

Christine Uber Grosse is president of SeaHarp Learning Solutions, a company that designs content for online cross-cultural training. She is professor emeritus of modern languages at Thunderbird School of Global Management. Her articles on business languages, online learning, and cross-cultural management have appeared in *The Modern Language Journal, Foreign Language Annals, Global Business Language, Journal of Language for International Business*, and *Business Communication Quarterly*. Her contribution, "Managing Communication within Virtual Intercultural Teams," was named Outstanding Article of the Year by *Business Communication Quarterly*. She served as president of Florida TESOL and president of TESOL's Video Interest Section. Christine divides her time among Mexico, Florida, and Arizona.

Maarit Valo

Maarit Valo is professor of speech communication in the Department of Communication, University of Jyväskylä in Finland. She served as head of the Department of Communication and as dean of the Faculty of Humanities. She supervises doctoral dissertations and teaches communication research methods and theory of mediated interpersonal communication in the graduate Speech Communication program. Valo founded the Section for Interpersonal Communication and Social Interaction in the European Communication Research and Education Association. Her research interests include technologically mediated communication, workplace communication, and communicative competence as part of professional expertise. She holds an adjunct professorship in the University of Helsinki in Finland, and she has been appointed twice as research fellow by the Academy of Finland. She serves as vice chair of the Union of University Professors in Finland. Valo was awarded her PhD from the University of Jyväskylä in Finland.

Michiel van Genuchten

Michiel van Genuchten, professor at Eindhoven University of Technology in the Netherlands, is manager of digital dentistry at Straumann in Switzerland. Previously, he was employed by Philips Electronics and ran his own software company. His studies are concerned with software as business, software management, and information technology support for virtual teams. Results of his research have been published in *IEEE Computer, IEEE Transactions on Professional Communication, IEEE Software, IEEE Transactions on Software Engineering,* and *Journal of MIS*. Van Genuchten received his PhD from Eindhoven University of Technology.

Doug Vogel

Doug Vogel, formerly at the University of Arizona, is now Chair Professor of Information Systems at City University of Hong Kong and an AIS Fellow. Prior to his faculty appointments, he was a systems analyst and developer and was head of his own company. Widely published, his studies include the integration of technology and education, with special emphasis on integrating audio, video, and data in interactive distributed education. Vogel is especially active in the development, facilitation, and evaluation of group support systems. He was awarded his PhD from the University of Minnesota.

Edward Volchok

After a long career as a marketing consultant, Edward Volchok joined the faculty of the business department at Queensborough Community College/CUNY in 2006. At Queensborough Community College, he chairs the faculty senate's Distance Education Committee. Volchok is also an adjunct professor at the Howe School of Technology Management's award-winning online graduate-level program at Stevens Institute of Technology. He developed Stevens Institute of Technology's WebCampus e-learn marketing management course in 2001 and has been the course instructor since it was first launched. Volchok earned his PhD from Columbia University.

FOREWORD

Internet-based online learning is a new phenomenon, originating in rudimentary form less than two decades ago and gradually building in scale and improving in quality and cost. Some of these advances are driven by a familiar story—the steady, unrelenting advances in the silicon chip, which has made computers and attendant software faster, lighter, more functional, and more reliable, and which, coupled with fiber cables, has produced revolutionary growth in communication.

Forward motion in Internet learning also has come from empirical knowledge on effective practices, drawn from tens of thousands of classes taught over the years. This book, by Robert Ubell and his excellent team of collaborators, adds an important dimension to effective teaching and learning in online environments. It addresses how interaction and collaboration online can be effectively harnessed in virtual teams. It is an important contribution to the larger field of Internet-based education.

To appreciate its usefulness, it is worth stepping back and reminding ourselves that to this day, Internet-based education has its fierce partisans and equally fierce detractors, although the latter have diminished in numbers and volume. Lost in the debate are essential characteristics that so markedly differentiate Internet learning from traditional "distance education," such as self-learning from books and other print media, correspondence courses, and television, all of which have coexisted for a long time. In fact, it is these often-forgotten methods that enable the introduction of virtual teams.

Internet-based education is quite different from previous styles of distance education. For the first time in history, we have a "distance education" that allows all the elements that we commonly associate with on-campus instruction—which continues to be viewed as the standard—to be available to distance learners. This advantage proves to be crucial in the success of virtual teams. If we reflect on the key resources associated with a high-quality campus education, we conclude there are basically three: (1) access to learning materials such as books, journals, and educational software; (2) availability of an instructor; and (3) communication with peer learners. Earlier, resources available to remote learners consisted largely of the first of these—learning materials—leaving the student with the daunting prospect of having to work alone to master the subject matter.

The chapters in this book explore cases derived from practice, encompassing virtual teams in specific fields, leadership, and team effectiveness, among other aspects, coupled with the technologies required for smooth interaction. The work covered here launches a critically important scholarly examination of effective virtual team practices. It is not an end, but rather, a beginning. It needs to be read by anyone with an interest in collaborative distance education. From these snapshots, practitioners will be able to draw their own conclusions about the future directions of virtual teaming.

Practices and ideas explored here should be of equal interest to many outside of academe. In modern society, one is hard pressed to think of useful products and services that can be developed effectively and delivered merely by a single person alone. Often vast teams of experts are needed, increasingly, which means participating in a virtual team with members likely to be scattered across several time zones. Because they operate largely asynchronously—creative interaction and collaboration can be carried out even with members in different times and at different places—teams can now carry out their responsibilities anytime and anyplace.

Virtual teaming is an important practice, and its importance and usefulness will only increase. Investigators who report their findings in this book have performed a valuable service by reminding us of the attributes offered by the Internet; perhaps the most significant of which is interaction and collaboration among people, now uniquely operating in virtual teams.

—Frank Mayadas
Alfred P. Sloan Foundation

PREFACE

One way to guarantee an informed, efficient workforce for the twenty-first century is through e-learning. Today, professionals can maintain their accreditation and expertise even when they live far from high-quality research universities.

In the United States, online enrollments have been growing steadily and the number of institutions offering e-learning programs has increased significantly. Viewing the trend from within the halls of an institution that serves students worldwide, I can confidently predict that the global demand for e-learning will surge as courses from topflight universities become available to students in emerging nations—especially those that lack the educational infrastructure, but not the desire, to educate their engineers, mathematicians, and scientists.

Polytechnic Institute of New York University's 155-year history is marked with discovery and innovation. So it is no surprise that we have taken enthusiastically to online learning. Our enrollment is expected to double over the next years and double again year after year. NYU-ePoly now offers 20 online high-tech and executive graduate degrees covering, among other subjects, clean energy, computer engineering, cyber security, and telecommunications and organizational behavior, with new programs in advanced technologies coming online nearly every semester.

Robert Ubell, head of NYU-ePoly, notes these pluses for distance learning:

- *It is global.* America's science and engineering schools are recognized leaders. Online curriculum gives access to qualified engineers, scientists, and managers everywhere.
- *It is good for the world.* Access to skills and education is essential for the growth and stability of all countries.
- *It assures continuing education.* Professionals in technical jobs need to update their skills as technology is transformed. E-learning is the most flexible way to guarantee employees stay current with the demands of today's—and tomorrow's—jobs.
- *It places the best students in the best universities.* Fortune 500 companies want to send employees to outstanding research universities because of the superior skills they acquire. High-ranking schools that offer online degrees provide leading companies with a strategically skilled workforce that matches their objectives, whether personnel are in Dubuque or Abu Dhabi.

A recent US Department of Education report, based on a decade of studies, should settle any lingering debate about the quality, benefits, or equivalence of e-learning. It found that "on average, students in online learning conditions performed better than those receiving face-to-face instruction."

Corporations increasingly perceive the importance of e-learning for employees. NYU-Poly has explicitly underwritten this relationship, and is working with respected corporations to match their corporate goals to the curricula we offer. Because professionals employed in scientific and technical fields require the rigorous academic standards offered by research universities, students must be confident that distance classes match the coursework and high academic standards offered in traditional classrooms.

E-learning is one of the keys to solving our global challenges, transforming our expectations, our employees, and the way we do business.

—Jerry M. Hultin
President, Polytechnic Institute of New York University

Dewey Goes Online

Robert Ubell

Nearly a century before the Internet entered college and university life with online learning, American philosopher and progressive education champion John Dewey recognized that traditional classrooms can often stand in the way of creative learning. Troubled by passive students in regimented rows, Dewey worried that docile students, accepting the unquestioned authority of teachers, not only undermined engaged learning but also thwarted democratic practice in the social and political life of the nation. Instead, Dewey called for a "spirit of free communication, of interchange of ideas" (Dewey, 1915, p. 11), encouraging "active, expressive" learning (Dewey, 1915, p. 20).

Taking up ideas suggested by Dewey and others,[1] progressive educators in the 1920s proposed that students learn best by performing real-life activities in collaboration with others. Experiential learning—"learning by doing"—coupled with problem solving and critical thinking, they claimed, is the key to dynamic knowledge acquisition. Rather than respect for

[1]Other early leaders of progressive education in the United States and abroad were American educator Francis Parker; German teacher Friedrich Fröbel, who coined the term "kindergarten"; Swiss school reformer Johann Heinrich Pestalozzi; Abraham Flexner, American medical-school reformer; and Johann Friedrich Herbart, German philosopher and psychologist who first introduced pedagogy as an academic discipline.

authority, they called for diversity, believing that students must be recognized for their individual talent, interests, and cultural identity.

Building on the work of Dewey and others, constructivist[2] ideas emerged in the 1970s and 1980s. Constructivists believed that knowledge is built on experience mediated by one's own prior knowledge and the experience of others, a philosophical tradition that goes back to Immanuel Kant. According to constructivists, learning is a socially adaptive process of assimilation, accommodation, and correction. For constructivists, students generate new knowledge on the foundation of previous learning.

In contrast, objectivists[3] believe that learning results from the passive transmission of information from instructor to student. For them, reception, not construction, is the key. Objectivists assume that reality is entirely open to observation, independent of our minds. Modern neuroscience appears to support the alternative constructivist claim, concluding that the brain is not a recording device, but rather, the mind actively constructs reality, with experience filtered through a cognitive framework of memories, expectations and emotions (Dehaene, 2002).

Progressive education was never widely embraced. Apart from a handful of elementary and high schools and a few colleges,[4] for the most part over the last century, schools

---

[2]Chief among constructivist theorists are American cognitive learning psychologist Jerome Bruner, Swiss developmental psychologist Jean Piaget, and early Soviet psychologist Lev Vigotsky.

[3]Principal objectivist theorists were the Russian (and later Soviet) psychologist Ivan Pavlov, famously known for his work on conditioned reflex in salivating dogs, and the American psychologist B. F. Skinner, who championed radical behaviorism in what he called operant conditioning.

[4]Among the handful of colleges and universities that continue the progressive education tradition are Bank Street College of Education, Goddard College, Antioch University, and Union Institute and University.

rejected progressive theories, preferring conventional practice instead, with students seated in rows facing the teacher, a scene reminiscent of turn-of-the-century vintage schoolroom photographs.

Face-to-face teaching, the most common style of instruction and, consequently, the practice that appears to be most natural, is often valorized as the foundation against which all other methods are measured (Russell, 2001). It is taken for granted that the classroom is the normal place for learning. Yet there is little evidence to support the claim that traditional education is the standard. The basic assumption is that face-to-face students form a cohesive group, participating alike in discussion, listening to lectures, building intellectual and social relationships with teachers and peers inside and outside class. But, as Anthony Picciano points out, this is not always the case. Classroom students often feel alienated, drawing away from others and isolating themselves (Picciano, 2002). A significant population feels estranged and falls into a pattern of failure.

Conventional education assumes that because students occupy the same space and are subject to the same conditions, they are fairly similar and should emerge with the same or similar learning outcomes, regardless of economic or social status. Because students are visible at their desks—rather than invisible in a virtual classroom—somehow we assume that we can know them and understand them. We believe that when we see students in physical space, we can actually gain access to them. Yet it's their invisible qualities that mostly determine who they are. According to Pierre Bourdieu (1989), we forget that the truth of any interaction is never captured entirely by observation. So while face-to-face interaction is often thought of as giving us perfect knowledge of student behavior, in fact, physical presence can often obscure crucially hidden social and psychological relations.

We tend to believe that visual cues—facial expressions and body language—offer us sufficient social communication markers to understand one another. Yet these actions, while open to inspection, fail to give us access to unseen psychological and status relationships to which we are often blind. The classroom resists distinctions that are formed by groups and hierarchies that crisscross it from outside. Traditional instruction—especially the classroom lecture—is a one-size-fits-all product that ignores student identities as multiple, overlapping constellations of real and imaginary selves.

What is visible can often be damaging, turning common experience against us. Hair style, clothes, our perceived ideas of physical beauty, and other personal characteristics can often undermine us, even as they have the capacity to move us closer together. The classroom is a place where ordinary misperceptions by teachers and students can easily defeat effective learning. It is a place where ethnicity, gender, and race are in plain sight, sadly subject to the same stereotypes and prejudices found in the streets. Online, however, students are often able to enter the virtual classroom anonymously, avoiding the stigmatization that can occur in physical space (Kassop, 2003).

Dewey raised his voice against the ordinary schoolroom, a place made almost exclusively "for listening." Following Dewey, Paulo Freire recognized the narrative character of the teacher–student relationship. "Education is suffering from a narration sickness," Freire (1970) observed and famously ridiculed conventional instruction for its "banking concept of education," with students mechanically memorizing content, turning them into instructional depositories.

Today, the demands of online learning—finding unprecedented ways to engage invisible students—have reclaimed Dewey. Suddenly, the lessons of progressive education and the constructivist legacy have become relevant. Rather than being discarded, Dewey is now seen as prescient. In one of the principal online learning research texts, Starr Roxanne Hiltz and her colleagues claim that collaborative online learning "is one of the most important implementations of the constructivist approach" (Hiltz and Goldman, 2005).

Constructivist strategies were introduced in online education and in virtual teams in industry to overcome what Karen Sobel Lojeski and Richard Reilly (2008) call "virtual distance," a consequence of a number of potentially alienating factors. Members of virtual teams are often widely separated geographically, with many located in distant time zones. Frequently composed of students from different cultures who work in different organizations, with unfamiliar standards and models of behavior, virtual teams may also consist of participants with varying technical proficiency.

According to Lojeski and Reilly, virtual distance is composed of three principal disturbances—physical, operational, and affinity distance, with physical distance emerging from obvious disparities in space and time. Operational distance, on the other hand, grows out of workplace dysfunction, such as communication failure—for example, receiving an e-mail from a colleague whose poorly articulated text cannot be deciphered. Affinity distance reflects emotional barriers that stand in the way of effective collaboration. Lojeski and Reilly claim that absence of affinity among team members is the greatest obstacle to quality performance. For them, reducing emotional estrangement in groups is the single most important task.

Pedagogy has never played a significant role in higher education. Instructors walk into most college classrooms without any special training in teaching skills. In universities, pedagogy is often dismissed as a discipline appropriate for kindergarten and elementary school, not a proper subject for higher education. With online learning, however, pedagogy emerges as a necessity. Without training in how to engage students, helping to close the online psychological gap, faculty are essentially unprepared to teach. In a turnaround, faculty now demand that they receive quality instruction about how to teach online before they enter their virtual classroom; otherwise, they feel stranded. For many, teaching online often requires wholesale reconsideration and reformulation of subject matter and delivery, a reassessment that can lead to rejuvenating faculty engagement and heighten the granularity of content.

Still, teaching online can be quite disorienting. Faculty can no longer rely on their ability to deliver performances that engage students intellectually and emotionally. In classrooms, professors practice many of the techniques employed by stage actors—rehearsal, scripting, improvisation, characterization, and stage presence (Pineau, 1994). Exploiting tension, timing, counterpoint, and humor with dramatic effect, skilled classroom teachers exhibit qualities that can stimulate thought and action. We are often drawn to content and energized by instructional performances.

But a practiced, smooth presentation can also hide the struggles that go into its creation. It can mask dislocations, errors, and false starts out of which lectures—and the multiple, contradictory acts of learning—are actually assembled. In the *Wizard of Oz*, when Toto pulls the curtain aside, the Wizard's booming, confident authority is revealed as merely manipulation by an ordinary man engineering his false self. According to French theorist Jacques Rancière (1991), the instructor's expert delivery may create deep fissures between student and teacher. While both may be physically in the same space in a classroom, faculty and student can be in far different places emotionally. The more skilled the lecture—often fascinating and pleasurable as in a stage performance—the more it may give the illusion that students have actually absorbed the lesson.

Unwittingly, the lecture contrasts the faculty's apparent confidence against the student's feelings of inadequacy. In a lecture, professors leave their uncertainty behind, submerging the battles they fought to generate a coherent narrative—struggles students are yet to face. In class, instructors present what is known at the very moment when students take their first steps into the unknown. Students soon discover that learning is a gradual, often stumbling, process that can lead down blind alleys, often hobbled by false starts. Marked by ruptures and dislocations, learning is a risk-taking exercise, not an elegant performance.

Online learning plays a part in a long trend that has unseated everything that was thought to be ineluctable, moving what was always thought to be at the center to the periphery. Copernicus drove the earth from the center of the universe to play merely a supporting role in a minor galaxy. Darwin displaced men and women as the pinnacle of creation, placing them as accidental creatures in a long evolutionary drama. While not as momentous, online learning, too, overturns conventional wisdom by drawing professors away from the front of the classroom and moving them to the side as observers (Ubell, 2004).

Still, while the authority of the faculty appears diminished in online teaching, their role is now no longer simply as a performer of narrative lessons. Online, they play a new part as complex agents of intellectual transformation. Merely assigning students to groups and encouraging them to work together will not yield results. Students are not automatically transformed into involved and thoughtful participants when they go online. Poorly prepared, peer learning can exacerbate status differences and generate dysfunctional interactions among students (Blumenfeld, 1996). At worst, virtual teaming can result in "the blind leading the blind" or "pooling ignorance" (Topping, 2005).

Faculty must orchestrate online learning, building "intellectual scaffolding," prompting students with projects, discussion topics, and questions to encourage them to think deeply, creatively, and interactively (Christudason, 2003). Ironically, moving from physical to online space often calls upon faculty to become far more engaged than in the classroom. Instructors become facilitators, propelling students to engage in discourse through discussion and argument to generate and link ideas.

At their best, online faculty achieve what is known as "teaching presence," a constellation of actions that give students a vivid sense that virtual instructors are fully engaged (Benbunan-Fich, 2005). Teaching presence emerges from online faculty–student interaction and feedback that exploits e-mail, chat, discussion boards, webinars, and other applications that defy the limits of space and time. Unlike the time constraints imposed by the physical classroom, online instructors and students enter a borderless space, open to the possibility of continuous dialogue (Kassop, 2003). In asynchronous communication, the give and take of online discussion is conducted at a much slower pace, giving students and teachers time to reflect, with more room for analysis, critique, and problem solving (Picciano, Online Learning, 2006).

Extending online learning to virtual teams, teaching presence recedes as peer-to-peer learning unfolds. Virtual teaming opens online space, allowing students to work together in pursuit of a shared goal or to produce a joint intellectual product. Student-to-student interaction in small groups permits the acquisition of knowledge and skill through collaborative help and support in what is known as "cognitive presence." For virtual teams to succeed, instructors must encourage students to practice collaborative skills—giving and receiving help, sharing and explaining content, and offering feedback, but also interrogation, critique, challenge, argument, and conflict.

With the teacher out of sight—whether online or on campus—student teams are lifted out of their seats and assume positions rarely taken before—as a leader, facilitator, reporter, observer, or participant (Swan, 2006). Ultimately, online students may achieve what is known as "social presence," a zone in which virtual teammates are not mere mental fictions but appear seemingly as "real." With virtual teaming, faculty release students from their paternalistic grip, freeing them from pedagogical infantilization, allowing them to find their own way as mature learners. At its best, virtual teaming emancipates students from the hierarchy of conventional education to practice intellectual democracy.

We can trace the history of education over the last decade by mapping the position of teachers as they migrate from the center of the educational stage as principal actors in traditional classrooms, move to the wings in online learning where they assume a supporting role, and finally depart in virtual teams, where they play an entirely new and radical part, setting the stage for students to act all on their own. Faculty now sit in the audience as observers and critics, with students on the platform as performers, occupying an engaged space where learning takes place collaboratively with their peers.

Teams disrupt the linear narrative of conventional instruction by introducing overlapping discourse, flowing from multiple sources in discontinuous, mostly asynchronous, peer-to-peer discussion and argument. In the spirit of Dewey, who encouraged learning by doing, the task of teams is to work together to create knowledge. For Dewey, the ideal classroom is a "social clearinghouse, where experiences and ideas are exchanged and subjected to criticism, where misconceptions are corrected, and new lines of thought and inquiry are set up" (Dewey, 1915, p. 34). Active learning, he claimed, emerges from students forming a "miniature community, an embryonic society" (Dewey, 1915, p. 13)—uncannily like virtual teams.

REFERENCES

Benbunan-Fich, Hiltz, R. and Harasim, L. (2005) The online interaction learning model. In: Starr Roxanne Hiltz and Ricki Goldman, editors. *Learning Together Online*. Mahwah, NJ: Lawrence Erlbaum, pp 28–29.

Blumenfeld, Phyllis, et al., (1996) Learning with peers, *Educational Researcher*, 25, pp 37-39.

Bourdieu, P. (1989) Social space and symbolic power. *Sociological Theory*, 7(1), 14–25.

Christudason, A. (2003) Peer learning. *Successful Learning.*, (37),

Dehaene, S. (2002) *The Cognitive Neuroscience of Consciousness*. Cambridge, MA: MIT.

Dewey, J. (1915) *The School and Society*. Chicago: Chicago University Press.

Freire, P. (1970) *Pedogogy of the Oppressed*. New York: Continuum.

Hiltz, S.R. and Goldman, R. (2005) *Learning Online Together*. Mahwah, NJ: Lawrence Erlbaum.

Kassop, M. (2003 May/June) Ten ways online education matches, or surpasses, face-to-face learning. *The Technology Source Archives*, 1–7.

Lojeski, K. S. and Reilly, R. (2008) *Uniting the Virtual Workforce*. Hoboken, NJ: John Wiley & Sons.

Picciano, A. (2002) Beyond student perceptions. *JALN*, 6(1), 21–39.

Pineau, E. L. (1994) Teaching is performance. *American Education Journal*, 31(1), 3–25.

Rancière, J. (1991) *The Ignorant Schoolmaster*. Stanford: Stanford University Press.

Russell, T. (2001) *The No Significant Difference Phenomenon (5th ed.)*. Montgomery, AL: IDECC.

Swan, Karen (2006) Assessment and collaboration in online learning. *JALN*, 10(1), 45–62.

Topping, K. (2005) Trends in peer learning. *Educational Psychology*, 25, 631–645.

Ubell, Robert and Mayadas, F. (2004) Online learning environments. In: A. DiStefano, K.E. Rudestam, and R.J. Silverman, editors, *Encyclopedia of Distributed Learning*. Thousand Oaks, CA: Sage.

PART

1

MANAGING VIRTUAL TEAMS

CHAPTER

1

BUILDING VIRTUAL TEAMS[1]

EDWARD VOLCHOK

Every instructor who teaches online should consider introducing team projects. While they are especially useful in marketing, virtual teams are a very effective educational tool in all disciplines. When Stevens Institute of Technology asked me to develop an online graduate marketing course, I was certain of one thing: I wanted to give my students an experience that would instill a deep appreciation of marketing and the challenges marketers face. To achieve my goal—to be truly hands-on—I knew I would have to assign team projects. The looming question was: How should I structure teams in a computer-mediated, asynchronous learning environment?

The online environment presents unique challenges, especially for team projects. Online, students and faculty lack a face-to-face connection on which we all depend in on-campus classes. Its importance cannot be understated. Without 30 hours of physical presence in a traditional classroom, students can feel isolated from their instructor as well as from one another.

[1]Based on an article in *eLearn* magazine, with permission.

Virtual Teamwork: Mastering the Art and Practice of Online Learning and Corporate Collaboration. Edited by Robert Ubell

Online relationships—built primarily by communicating with the written word, without benefit of body language, a passing smile, or an occasional joke—can seem hollow. The lack of these cues makes building trust very difficult. The wise use of virtual teams, however, can help overcome virtual distance.

Since early 2002, I have been teaching "Marketing Management," an online graduate course in Stevens Institute of Technology e-learning unit. From the start, I formed virtual teams. Having seen the triumphs and tribulations of over 75 teams on some 300 team projects in 20 classes, I will discuss some of the methods I use—and that you might also employ—to make your virtual teams succeed.

STRUCTURING ONLINE CLASSES

My primary task was to migrate the school's "Marketing Management" course from the classroom to the online environment. To keep virtual students motivated, I want them to wrestle with real marketing problems. I do not encourage students, already isolated from classmates and their instructor, to cram alone for an exam and spout back passages from the textbook, only to forget key lessons once the semester ends. My course is designed to make students confront the challenges marketers face; they learn by doing. Students get their hands dirty by tackling real marketing problems. Merely studying for tests is just not a sufficiently involving experience. I banished tests.

My next challenge was the lecture. I enjoy lecturing and I have given well-received lectures in the marketing classes I taught at New York University. I could have easily adapted them and added a few new ones. As they are already mounted on PowerPoint slides, posting them on the course web site would have been easy. While I have been a fan of PowerPoint since 1985, when I beta-tested its very first release, PowerPoint is *not* a robust teaching tool for online classes. To be effective, PowerPoint slides need a parallel real-time presenter. While I could have recorded an audio track to accompany my slides—some instructors report great success doing just that and I have heard students praise the experience—prerecorded lectures are just *not* interactive.

It became clear that to build a successful online course, I would have to structure my classes around case studies, not lectures. So I introduced two types of case studies. The first is a series of nine small cases designed to be solved individually. These briefs focus on a single topic. Students are given one week to develop, present, and discuss the assignment. The second—the showpiece of the course—is an in-depth case study. These cases require students to work in virtual teams to solve real strategic and marketing plan objectives. The teams have two weeks to present and discuss their cases. As preparation often takes longer, teams may be given extra time to prepare their projects, some until the end of the semester.

Students post their individual and team projects for class discussion on a message board on the course web site. Besides coaching the teams as they prepare their projects, the principal way I insert myself in the process is by giving teams due dates and, at the end of each project, preparing an overview of the assignment, recapping strengths and weaknesses of solutions presented, and resolving issues that may have arisen during discussions. In the end, I post grades.

STUDENTS AND MARKETING MYTHS

Nearly all my students are working professionals. Typically, they may have already earned a degree in engineering or in another technological discipline. Their employers encourage them to get advanced degrees and most companies pick up the tab for all or part of their tuition. Employers are top corporations, such as Verizon, Pearson, JPMorgan Chase, Pfizer, Boeing, Honeywell, and Johnson & Johnson.

For working professionals, online courses with anytime, anywhere opportunity offer important benefits. Online education enables them to complete their degrees faster and more conveniently than if they attended traditional classes on campus. Hectic and ever-changing schedules, pressing demands of business travel, and the obligations of growing families make attending a fixed schedule of on-campus classes inconvenient, if not impossible.

Today's best marketers say that marketing is too important to be left to their department only. That's why good teamwork is essential for successful marketing. In today's competitive marketplace, businesses must focus their entire organization on delivering superior value to customers. In practice, this means building cross-functional teams. The marketing department can no longer dictate objectives, strategies, and timetables. It must engage other departments actively and seek their support. Marketers must earn trust, which requires teamwork. For courses designed to give students realistic experiences wrestling with delivering customer value, team projects are essential. Through them, students learn the art of debating, generating consensus, and delivering cogent proposals under tight deadlines. To be collaborative team players, students need to learn how to lead *and* follow.

In my consulting practice, I have observed clients—multinationals and start-ups—increasingly rely on collaborative work performed by computer-mediated teams. As Robert Ubell, editor of this book, says, "Virtual teams replicate the way industry, commerce, and research is practiced everyday worldwide." Virtual teams not only are appropriate for presenting marketing subject matter in its proper context and meeting the needs of my time-starved students, but also enable students to master unique challenges of participating in virtual teams in a relatively risk-free university environment.

STUDENT RESPONSE TO VIRTUAL TEAMS

Students have strong—and conflicting—opinions about virtual teams. Most love *and* hate their team. Even students in a well-run team are frequently frustrated. At times, they would gladly abandon their project or, at the very least, a teammate or two. Students who share their frustrations with me about getting teammates to pull their weight remind me of the famous line, "Hell is other people," in Jean-Paul Sartre's play, *No Exit*. There is an important lesson here. Teamwork is not easy. Teammates often have unspoken agendas, which may not parallel the team's. Getting a virtual team to work together can often seem as vexing as herding cats. Yet we need not echo Sartre's pouting pessimism. You can develop enthusiastic teamwork in an online environment. Instructors can give teams boosts by acting as mentor, psychologist, rabbi, and arbitrator.

Suggestions that follow are designed to make virtual teams less like the strident interpersonal transactions from a French existentialist's play and more like the immortal 1927 New York Yankees.

Tip 1. Get teams off to a strong start. To get your teams off to a running start, you must set clear expectations. My online class syllabus, for instance, is far more detailed than those in traditional courses. It's wise to post messages about your expectations in several places—in assignments, throughout the course site, in periodic class e-mails, and in detailed reviews of each project.

It's worthwhile to get to know your students well. In my orientation survey, I ask: What do you do for a living? How much work experience do you have? What did you study as an undergraduate? What degree are you studying for now? What do you know about marketing and marketers? In a traditional classroom, this information is gathered face-to-face. Obviously, online, such meetings are impossible. I am in New York City, while my students are all around the globe—Maryland, Illinois, California, Hong Kong, Taipei, and Moscow. Even for those within commuting distance to our Hoboken campus, tight schedules make face-to-face meetings difficult.

In addition to the survey, I hold a 10–15-minute phone conversation with each student a week before our online class begins. My purpose is to greet students, tell them about myself, clarify what I expect, and hear them express what they hope to achieve. These conversations are critical for establishing teacher presence in virtual classes.

It's also useful to give students a chance to introduce themselves to each other. My first individual assignment is designed to build strong teams. I ask students to post a personal statement, sometimes called a "two-minute pitch" or "elevator pitch." I require the assignment for two reasons. First, students will need to perfect pitches in their careers. Helping them market themselves is a good way to introduce them to the field. Second, from the team's perspective, students need to know their teammates and their competition.

Tip 2. Establish teams in the first week. To give teams a running start on their first team project, which is due the fourth week of class, I announce team rosters at the start of the first week. Typically, my teams are composed of three to five members. Because I assign four team projects, I aim for four students per team. I recommend that students not attempt to form their own teams. Virtual teams are not like pickup teams at the playground. Employees in a company do not commonly select their teammates. Besides, students do not have the time or the information necessary to form teams on their own. To move things along, instructors should select team members by trying to balance experience, skills, and background.

Tip 3. Contain the "Mussolini." A good team player must act as both a leader and a follower. It's best not to allow team members to dominate a team by force of personality. To preempt dictators, I recommend rotating team captains. With four members on each team and four team projects, every member will rotate leader and follower roles.

Team captains have critical responsibilities. They set the agenda, distribute assignments, and enforce deadlines. The most important responsibility of the captain is to make sure the team presents a unified solution to a problem. To achieve it, the captain must bring conflicting points of view to the fore and achieve consensus. A good team captain is a consensus builder, not a despot. Wise captains use their persuasive powers to harmonize the team's ideas. They also make certain that contributions of individual team members are consistent with the team's objectives and strategies so that the final submission is presented in a unified voice. If a team submits its project with disparate sections, each written in a different style, the team lacks the essential cohesion of a successful team. With an ineffectual captain, disorder reigns. In such cases, the team experience can quickly sour.

Some students may cast themselves in the role of a "Mussolini." They force their views on others and try to stifle discussion. When I receive e-mail messages complaining that a student is trying to suppress the opinions of teammates and force the team to accept only his or her ideas, I telephone the self-proclaimed dictator. A few gentle reminders about team spirit generally work wonders. Most students respond well to appropriate intervention.

Tip 4. Empower "Shrinking Violets," restrain "Rambos." Worse than "Mussolinis" are "Shrinking Violets" and go-it-alone "Rambos." In one of my online classes, a "Shrinking Violet" was lurking during the first team project. Her teammates complained that this wallflower was not contributing anything, disappearing entirely from the second team project. I telephoned the student to warn her that lack of participation in team projects may result in failing the course and wondered whether she wished to drop the class. While she assured me that she hoped to complete the course and that her performance would improve—which it did for a week—she then vanished entirely, reappearing 20 hours before the next project was due. She frantically e-mailed her teammates, begging for something to do. Her teammates responded politely, saying, in essence, "Thanks, but no thanks." Violet then morphed into "Rambo," privately e-mailing me her project two days late, with a fusillade of hostile remarks directed toward her teammates. A quick glance at her submission led me to suspect that she had incorporated her team's research—posted on the teams' discussion board—into her own project. Because I had been observing the team's message board, I knew she had not contributed. I reminded her that I do not accept late assignments and, what's more, I might enforce the university's policy on academic integrity, which carries severe penalties for students who "borrow" the work of others. Ultimately, she accepted a failing grade for the assignment.

Establishing a strict team structure may help reduce such incidents and the need to take serious corrective action. Still, if the structure fails, instructors must hold feet to the fire. You must persuade and cajole "Shrinking Violets" to participate fully. You must encourage "Rambos" to work with their teammates. Ply them with carrots, but if all else fails, beat them with sticks.

Tip 5. Give students tools to communicate. Students should have as many tools as possible to communicate with teammates. When I announce the composition of teams, I distribute the team's e-mail addresses and telephone numbers. I also establish team-specific synchronous chat rooms in which conversations are automatically archived, giving all members a record of discussions. I also provide each team with a private asynchronous message board. Most teams make extensive use of the board to post drafts, suggest revisions, and reconcile opposing views.

I then offer suggestions on how best to work together. I soon step aside to give teams room to succeed or fail. Students are very inventive in finding ways to work together. Sometimes, they use their company's conferencing systems. Occasionally, teams go to great lengths to meet in person. While this is rare, many have told me of their intense desire to meet their teammates face-to-face.

Tip 6. Enlist students in holding teammates accountable. The most unfortunate consequence of poor teaming is when students ride on the coattails of others. However, there are a number of ways of preventing those who try to get a free ride. Consider these two: It's prudent to ask students in a team to assess everyone else's contribution to the project. This requirement encourages them to participate actively because they are aware that I will take these evaluations into account when determining their final grade.

Under certain conditions, however, teams may vigorously defend those who fail to fully participate. Occasionally, a team will urge instructors to give the same grade to those who did not contribute actively because the student is viewed as a valuable partner, even if he or she may not have been as engaged. You may learn that "Robert did not fully participate because his wife just gave birth prematurely," or "Maria did not fully contribute because she suffered minor injuries in a car accident." You can easily confirm these accounts with a simple phone call, and if you discover they are true, it's wise to follow the team's recommendation. Support for a teammate in need shows that the team has built an *esprit de corps*, a quality that deserves encouragement.

Another way of helping to generate participation in a team project is to give a separate grade for discussion. Discussions are based in part on questions I pose to each team and questions students pose to members of other teams. How students respond individually is an excellent way to determine how deeply they understand the assignment.

Tip 7. Encourage a competitive spirit. Encourage teams to compete against each other. In my online classes, teams work on the same case and I am amazed at the variety of solutions competing teams offer. After they post solutions on the discussion board, a vibrant discussion ensues. To get the discussion started, I recommend that the instructor post questions about the solutions. While some of my questions are legitimate, others are deliberately tricky, the kind that a confused or hostile boss might ask. Divergent views among different teams give you an opportunity to motivate students to dig deeper into the subject. For example, in response to one assignment, a team might argue that the solution to a particular problem is to raise prices. Another might recommend cutting or holding prices. In this situation, you might ask team members to comment on why their solution is better than the other. The debate requires students to delve into other presentations, creating a lively discussion. Then I post my project summary in which I outline key concepts, reconcile or refute alternative solutions, and offer suggestions for the next team project.

Tip 8. Reward risk-takers. Encourage students to think big and take risks. I often tell students, "If you are going to make a mistake, make a big one." In my virtual classes, big mistakes come with few penalties, especially if students can justify their ideas with sound arguments. Student errors do not lead to lower stock prices, the loss of millions of dollars, closed plants, or ruined careers. They do, however, open the class to new ideas and lively discussion. And that's what makes a winning course.

Tip 9. Unless asked, do not participate in team discussions. While it is very important for instructors to provide strong teacher presence in online classes, over-involvement can cause students to stop "talking." Give clear assignments, and then stand back to let students become engaged by themselves. Allow teams to make their own mistakes. Not entering the discussion does not mean you should keep silent altogether. When asked, it's best to respond promptly. When students request help or seek clarification, be generous with your time. It's wise to provide a detailed summary of every project before the next one starts. In my virtual classes, discussions close the day before the next module begins. I post my summary, highlighting the best and least successful team presentations. Then I post grades.

Tip 10. Consider real-time presentations. Yes, I know that live presentations violate the *anytime, anywhere* rule in an asynchronous course. But as Ralph Waldo Emerson said, "A foolish consistency is the hobgoblin of little minds." Stevens uses a webcasting software application designed for real-time presentations. Through web conferencing, teams engage in live conversations with voice over PowerPoint slides, presenting their solutions and responding to questions. Students love these sessions and so do I. They are as interactive and engaging as in any traditional classroom. They help overcome the isolation of distance education, with students feeling the presence of their instructor as well as their classmates. Students unable to attend webcasting sessions, owing to work and family constraints, can retrieve archived versions, complete with images and audio.

Nothing Succeeds Like Failure

Let's face facts. Not every team hits a home run. In fact, some teams strike out, even as good pitches fly by. We all learn a lot from our mistakes. Failure, after all, is a great learning experience. When my students submit a less-than-adequate team project, some may echo one of Sartre's more pessimistic and misanthropic sentiments: teams are "a useless passion." In the end, most consult with their teammates to improve coordination, planning, and presentation so that their next effort scores. This is, of course, the most important lesson any instructor can offer.

CHAPTER

2

LEADERSHIP

MICHAEL R. RYAN

The crucial element that differentiates a virtual team from a traditional team is the lack of face-to-face engagements. But is that the only differentiator? In order to be considered a team—and not merely a group—there must be a degree of interdependence. Without interdependence, it is simply a workgroup with output as the collective sum of individual efforts.

Some limit virtual teams to those that are globally or geographically dispersed, even suggesting that members must represent at least two nations. Consequently, they often use the term "globally distributed," rather than "virtual." Others say that members of virtual teams must communicate with one another in technologically assisted ways. For our purposes, we have identified three elements as essential to the definition of virtual teams:

> A virtual team (1) requires little or no face-to-face interaction and is dispersed geographically, organizationally, socially, or culturally; (2) its members communicate with each other in a technologically facilitated mode; and (3) members often communicate with one another asynchronously.

Virtual Teamwork: Mastering the Art and Practice of Online Learning and Corporate Collaboration. Edited by Robert Ubell

Virtual teams have been adopted for many reasons, but principally because of their obvious technological and economic advantages. They have also been introduced for other cogent reasons—to enhance diversity, to engage human capital more effectively, and to pursue dynamic market possibilities.

The ubiquitous computer, accompanied by dynamic changes in telecommunications, has turned the world into a global neighborhood. Technology permits employees to reach across the globe instantly and simultaneously and enables members of far-flung teams to work together far more collaboratively than ever before.

Economic benefits have emerged as a direct result of the introduction of powerful advanced technologies. Today, organizations can establish sophisticated communication channels by exploiting vastly superior new technologies, frequently requiring minimum investment. Since virtual teams often take advantage of infrastructure already deployed for information technology, advanced communications reduce the need for expensive travel as well as loss of productivity owing to extended travel time.

With the growth of virtual teams, organization can easily diversify. Once physical constraints are removed, teams can span multiple geopolitical and socioeconomic communities. Diversity enhances team richness by bringing in a wider range of perspectives. Recognizing the centrality of human capital, organizations are taking steps to increase the effectiveness of experts in decision-making. Before the introduction of virtual teams, availability of experts of crucial importance to the success of projects was often severely limited by geography. Today, knowledgeable experts may be engaged by multiple teams at the same time without ever leaving their normal workplace.

With increased globalization of the marketplace, virtual teams have emerged as a strategic corporate initiative. Virtual teams support decentralization, allowing organizations to establish a presence almost anywhere. Members of widely dispersed operations can now communicate with headquarters routinely, allowing distant personnel to engage significantly in core business activities and participate in strategic decisions. Members of global operations can now act as a single community.

While organizations can gain significant competitive advantage by deploying virtual teams, nevertheless there are potential obstacles. What issues must be addressed in order to succeed? One of the most obvious issues is how effectively are virtual teams formed? Clearly, success depends crucially on facilitating interdependence and a shared mental model among members.

Interdependence in its most basic form requires two conditions. First, the by-product or output of one member must be required by another to complete the task. Second, members must reach an agreement to exchange resources needed; the first member must be willing to provide the resource and the second member must accept it. In practice, interdependence often involves multiple levels of exchanges and consequently serves to reinforce the social network within the team.

As a result, a shared mental model emerges with greater cohesion and involvement among members. Cohesion is built on a higher level of commitment from each member. Rather than a passive acceptance of someone else's vision and goals, a shared mental model indicates that the member has actively adopted the vision and goals of the team as his own. This extended body of owners does not dilute the value of the vision, but instead increases the number of parties that have a vested interest in ensuring success of the team's goals. A result of an individual's greater commitment and shared mental model is that members will accept a more demanding role when the situation calls for it. Finally, as members commit to the shared mental model, they hold themselves accountable to achieving the team's vision and goals. Accountability promotes the team's success. It also serves as another unifying factor, drawing members into an ever greater cohesive force.

VIRTUAL DISTANCE

Members of a virtual team enter their environment under several constraints. Since "virtualness" exists along a continuum, it is influenced by a number of factors. As Sobel Lojeski et al. (2006) report, the continuum is a measure of Virtual Distance™ exhibiting these characteristics:

Relational Histories. Prior relational histories may involve individuals or groups. For individuals, personal histories may reflect direct or indirect current or previous relationships. In an indirect relationship, two members of a team may both have had—or continue to have—a relationship with a third participant or another person outside the team. In groups, relational histories may result from prior or present participation in a functional or corporate relationship. Such relationships may either cause difficulties or benefit the team.

Cultural Factors. Cultural factors may emerge from socioeconomic, racial, religious, corporate, or other culturally diverse perspectives. A particularly charged cultural factor derives from social order. Hierarchies commonly exist within teams, whether they are commercial, academic, or governmental. The difficulty occurs when social perceptions impede communication, especially when members believe that their cultural values dictate that status is the result of natural order (Hampden-Turner and Trompenaars, 1993).

Infrastructure. Some organizations may not be equipped with robust communication tools and other technologically enhanced infrastructure to support team members effectively. Should underlying support be poor, the team may fail to meet its objectives.

Isolation. Separation among team members, as well as between individual members and their supporting environment, may contribute to a sense of isolation. The loss of the "water cooler" effect (Jones et al., 2005) may deny members their casual, off-line conversations that often occur around a water cooler, lunchroom, or hallway. While they may have little direct relevance to work product, casual interactions often serve to produce or strengthen ties (see the section on Social Network below), contributing to building a greater level of trust between participants.

Task Interdependence. Greater interdependence between members decreases the perception of distance among them.

Team Size. The larger the team, the more likely that subgroups will emerge. Subgroups may challenge a shared mental model needed for a team to overcome difference.

Face-to-Face Interaction. The frequency and quality of face-to-face interaction can serve to either reduce or promote perceived distance. In face-to-face relationships, communication is aided by the presence of visible social cues. When social cues are missing, further constraints are placed on virtual teams. However, the relative anonymity of team environments can reduce inhibitions among members (Straus and McGrath, 1994; Cappel and Windsor, 2000; Martins et al., 2004). There is evidence that reduced inhibitions may also result from a lack of observable social or cultural differences in face-to-face environments that can generate obstacles created by hierarchical social status (Schmidt et al., 2001; Martins et al., 2004).

Multitasking. Today, it is the rule, rather than the exception, that employees are involved in multiple activities simultaneously. The greater the demands on the team from outside, the greater will be the separation from the team.

Technical Skills. When members are challenged by lack of technical facility with the demands of their project or are unfamiliar with the communication tools, relationships are likely to be inhibited. Team members who lack critical technological skills may find themselves isolated.

Collectively, these constraints determine the level of virtualness within a team. They also provide focal points for the leader attempting to unify the team to pursue collective tasks.

The Social Network

Social network analysis is concerned with the relationships among individuals and groups. Even though virtual teams are not actively engaged in face-to-face interaction, nonetheless they are social networks. In a virtual team, relationships may be direct or indirect. They may also have varying degrees of relative strength. Since we are by nature social beings, in any social situation there is a natural tendency to congregate with those with whom we share the greatest affinity. The process of joining others with shared characteristics, traits, interests, and so on serves to reduce the uncertainty presented by any novel environment (Fiol and O'Connor, 2005). But shared characteristics may also introduce faultlines (Lau and Murnighan, 1998) separating one group from others. The more the similarities shared by individuals within a group, the stronger the faultline.

While traditional teams frequently divide along functional lines, virtual teams present additional complexities. Faultlines that may develop in virtual environments can be magnified because the distances from others are more pronounced. While functional differences may exist, they are often secondary to geographical, social, cultural, or other differences. Faultlines that may divide the team into multiple subgroups—each with its unique social identity—present significant challenges to fulfilling the team's purpose, presenting serious obstacles for the team's leader. Conversely, if managed effectively, the challenge may present a significant opportunity.

In social networks, there are both individuals within a network and an aggregate of individuals, or a subgroup, within a network. In social network analysis, an individual is often termed a node. Likewise, a subgroup may also be referred to as a node. In relationships between nodes, whether at the individual or subgroup level, it's important to consider the strength of ties between nodes. A tie is the cohesion that exists between two nodes. The more the properties shared by nodes, the stronger the tie. This often leads to an obvious assumption that strong ties are essential to create a cohesive team. Paradoxically, strong ties may impede the team in pursuit of its goals. As the number of shared properties grows, disparity between nodes is reduced. This in turn reduces the body of knowledge and therefore the potential to uncover new ways of thinking to support the group's goals. Consequently, the most efficient social networks are often composed of nodes, both as individuals and as subgroups that exhibit weaker ties. Teams with weak ties are likely to bring greater knowledge and diversity to bear in fulfilling the group's objectives.

In a subgroup, shared attributes may also indicate a set of shared norms and behaviors. Social identity is often maintained by adopting shared norms and behaviors as one's own (Hogg and Terry, 2001). Adoption of a subgroup's norms and behaviors may create potential conflict with those of the collective team. Virtual team leaders must not only be aware of this threat but also act to defend the cohesion of the larger group.

The challenge in a virtual team is to develop a new allegiance to the larger team and, eventually, to instill an identity with the greater collective unit without compelling individual members to abandon their subgroup identity. Ideally, a new, collective identity will then coexist or complement one's subgroup. In some cases, collective group identity may threaten the subgroup. When the broader identity of the virtual team threatens a member's subgroup, it can often be detrimental to the team's cohesiveness. Virtual team allegiance can be undermined even more severely if many subgroups in the team also feel threatened. The virtual team leader is then placed in the unenviable position of having to address multiple conflicts concurrently in order to establish the legitimacy of the virtual team, restoring or developing a cohesive team identity.

AMBASSADORIAL LEADERSHIP

How—given the many factors that contribute to distance between members of a virtual team—can team leaders surmount these obstacles? As a solution, we propose a series of behaviors aimed at reducing virtual distance by addressing some of the social networking and cultural and social hurtles that contribute to perceived distance.

Ambassadorial Leadership is not presented as an extensive list of all leadership behaviors that contribute to successful virtual teams. Rather, it is a set of approaches that complement prevailing leadership models. These proposed behaviors emerged from a study of the challenges presented by virtual teams and are offered as a means to encourage a more collaborative environment. Ambassadorial Leadership supports these behaviors:

1. Internal boundary spanning
2. External boundary spanning
3. Shared/delegated leadership
4. Recognition
5. Advocacy

Let's take a close look at leadership behaviors designed to contribute to the resolution of constraints inherent in virtual teams; but at the same time, let's not forget the role played by individual team members. To understand how these behaviors may contribute to the overall effectiveness of a team, we must appreciate how members become effectively engaged in their team's goals and, consequently, how they can act to propel the team forward. While team leaders may introduce specific initiatives, members themselves must complete the tasks set out before them if the team is to be successful.

TEAM MEMBER BEHAVIOR

No matter how resourceful the leader is, if team members do not recognize their own value and adopt a corresponding set of behaviors and norms as positive standards, the team will be thwarted. Team leaders must create an environment that demonstrates support for each individual member. This in turn provides the foundation for members to develop greater cohesion with the group based on this perceived organizational support (POS). A complimentary behavior that frequently results from POS providing an indication of each member's commitment to the team is organizational citizenship behavior (OCB).

Perceived Organizational Support

Perceived organizational support is achieved when "employees in an organization form global beliefs concerning the extent to which the organization values their contributions and cares about their well-being" (Eisenberger et al., 1986). This early definition was coined to capture the relationship between members and their organization that exceeded those generally accepted at the time. Earlier organizational research considered only the economic exchange between employee and organization (March and Simon, 1958; Gould, 1979) or the emotional exchange (Levinson, 1965; Buchanan, 1974). POS acknowledges the importance of employees' perception of the value the organization places on their efforts. Assessment of the strength of POS occurs at the individual level, reflecting the commitment an individual exhibits toward his or her organization. Commitment can be viewed at three levels—continuance, normative, and affective.

Continuance commitment is an economic exchange between the individual and the organization, seen as the worker's "need to" maintain the relationship for essentially economic reasons.

Normative commitment is the next level in which the individual perceives an obligation to the organization. Such a commitment may exist for any of a number of reasons, but it results in an obligation that the individual "ought to" maintain.

Affective commitment represents the greatest personal allegiance in which employees are committed to the organization for emotional reasons. It is a commitment that reflects an alignment between one's personal beliefs and norms and those of the organization. Under an affective commitment, workers "want to" maintain their ties to the organization.

Positive POS goes beyond economic benefits, representing a greater effort on the part of the employee to contribute to the success of the organization (Meyer and Allen, 1991; Shore and Tetrick, 1991; Meyer et al., 1993; Allen and Meyer, 1996). While still subject to an exchange, POS recognizes the belief among certain workers that their efforts on behalf of the organization will be rewarded, not only financially but also with recognition and approval (March and Simon, 1958; Gould, 1979; Eisenberger et al., 1986, 1990; Wayne et al., 1997, 2002). When organizations acknowledge the contributions of their employees fairly and genuinely, workers are likely to exhibit positive POS (Blau, 1964). Recognition of value gives employees a sense of belonging (Eisenberger et al., 1990) and satisfies their need for praise and approval.

Minimally, the exchange is transactional (March and Simon, 1958), with the next level being an exchange based on a perception of fairness. In an expanded third level, Cardona et al. (2004) added work exchange in which the employee achieves intangible benefits such as variety, autonomy, and identity. Such exchanges demonstrate a greater discretionary contribution to the welfare of the organization and exhibit a greater level of POS (Shore and Shore, 1995). Organizations that engage employees at the highest level, promoting impressive levels of POS, are most likely to secure the greatest commitment from their personnel.

ORGANIZATIONAL CITIZENSHIP BEHAVIOR

In 1988, Organ proposed that organizational citizenship behavior "represents individual behavior that is discretionary, not directly or explicitly recognized by the formal reward system, and that in the aggregate promotes the effective functioning of the organization" (Organ, 1988, p. 4). As originally formulated,

OCB exhibited five primary behaviors—altruism, courtesy, conscientiousness, sportsmanship, and civic virtue.

Altruism is the willingness of an individual to assist another when the need exists.

Courtesy moderates or alleviates problems for others in the future.

Conscientiousness represents efforts that surpass an individual's role requirements.

Sportsmanship allows us to overlook minor disturbances and shortcomings of others.

Civic virtue refers to an individual's involvement within the political life of the organization (Deluga, 1994, 1998; Podsakoff et al., 1997). From the team's perspective, it is obvious how these behaviors contribute to collective effectiveness.

OCB is highly correlated with job satisfaction (Organ, 1988; Kidwell et al., 1997) and group performance (Organ, 1988; Podsakoff et al., 1997; MacKenzie et al., 2001). As suggested by Cardona et al. (2004), "People who perceive the relationship as a fair social exchange tend to increase their attachment to the organization, and this increased attachment encourages OCB. People who perceive the relationship as an unfair social exchange tend to decrease their attachment to the organization, to redefine the relationship as an economic exchange, and to limit further activities accordingly."

OCB is discretionary (Organ, 1988) and goes beyond the normal scope of an employee's responsibility. It provides measurable benefits to the organization (Kanjorski and Pugh, 1994). The gain from performance is thought to be the result of better use of limited resources as well as enhanced productivity (Organ, 1988; MacKenzie et al., 1993).

Clearly, there is a mediating relationship between POS and OCB. Undoubtedly, the individual's commitment can contribute significantly to OCB. If a team leader can effectively promote a shared mental model that results in alignment of beliefs and norms, members will tend to be committed at the affective level and therefore more likely to contribute discretionary zeal.

LEADERSHIP BEHAVIORS

INTERNAL BOUNDARY SPANNING

Virtual teams present a unique challenge. Unlike traditional teams, where boundaries exist primarily between the team and external entities, virtual teams are also faced with internal boundaries formed by real or symbolic barriers to access or transfer of information, goods, or people (Katz and Kahn, 1978).

In a virtual team, internal boundary spanning bridges faultlines within a diverse team. The pressure to satisfy needs of the group is the driving force that creates interdependence between team members and dictates the intensity of boundary spanning. The more diverse the team, the greater the need to exchange information actively between members. For both individuals and subgroups, internal boundary spanning can overcome physical, social, or psychological barriers. As the team engages more collaboratively, members tend to perceive that the group is acting more effectively, eventually contributing to team cohesion (Cohen et al., 1996).

EXTERNAL BOUNDARY SPANNING

Teams—whether traditional face-to-face or virtual—do not exist in a vacuum. Part of their effectiveness hinges on the relationship between the team and its external sources.

To overcome external boundaries, teams face a variety of outside constraints—resource acquisition, information gathering and feedback, and communication channels (Ancona and Caldwell, 1992a). Long before present-day virtual teaming research, Everett Rogers, in his 1960 work, *Diffusion of Innovations*, addressed outside obstacles from a social networking perspective.

Resource acquisition covers the ability of the team to secure resources that may not exist internally to perform its tasks, including time, finances, information, or even additional team members.

Information gathering and feedback identifies perceived value of the team's offerings, aiming to ensure the team parallels with the needs of its outside stakeholders.

Communication channels act to release information the team deems important. These are also used as vehicles to generate recognition and advocacy.

Exchange of information between the team and its external stakeholders often rests on the shoulders of a team representative, not necessarily the team leader. This is particularly true in virtual team environments where external entities are spread over many more domains—physical, political, functional, and so on—than the traditional team. From a social network perspective, the team representative may serve as gatekeeper or broker. Team representatives attempt to bridge the gap—called a "structural hole" by Burt (1995, 2004)—between the team and its outside constituents. In their gatekeeper's role, they monitor information from the outside and serve as a conduit bringing results back to the team. As brokers, they represent the team and release information as appropriate to support the team's objectives.

External boundary spanning allows the team to escape the limits of its own infrastructure, introducing new concepts from outside. Through spanning, the team monitors the external environment to ensure that its goals remain aligned with its parent organization.

SHARED LEADERSHIP

Shared leadership in teams is not a new concept. In earlier research, it was called, in addition to "shared leadership," "distributed leadership," "collective leadership," and "peer leadership." The concept first appeared in the work of the turn-of-the-century scholar, Mary Parker Follett, a pioneer in the qualitative study of organizations. In her 1927 conference series (Follett, 1927/1942, p. 249), she proposed that shared leadership is *a fait accompli*, "... a system based neither on equality nor on arbitrary authority, but on functional unity" (1942).

Shared leadership and delegation of authority is often based on the situation as well as skill sets needed by the group. It recognizes that in a diverse and dispersed team, a single individual may not be able to fulfill all the leadership roles that may be needed over the life of a team, especially within the subgroups' emerging functional faultlines. Shared leadership confers status and responsibility on selected team members, often resulting in stronger team cohesion.

Experience tells us that there is more to leadership than simply formal authority. Consequently, it is important to understand how we define the leader. A leader may be designated by their formal position—those who are given positional power—or alternately, leaders may emerge from some other mode of influence. Those who possess referent or expert power may emerge as key figures in a leaderless, self-managed, or self-directed workgroup. Shared leadership may exist in a variety of forms. A leader may recognize his or her shortcomings overtly and bestow authority on others (Pearce and Sims, 2002). Shared leadership may also emerge in less direct ways. In the absence of a lead actor, another may assume a leadership position until the team reaches its desired outcome or a previously assigned lead actor may return and assume his or her position once again.

Shared leadership in virtual teams may emerge naturally, particularly when subgroups find themselves physically, socially, or psychologically distant from an assigned team leader. Likewise, the social system within a subunit may equally restrict the emergence of shared leadership. In these situations, team leaders must actively promote shared leadership. By promoting shared leadership, virtual team leaders encourage a more active exchange between team members and, just as important, allow peer-to-peer influence to deepen the team's qualities.

Recognition

Recognition contributes to cohesion. It reinforces feelings of self-worth of individual team members and subgroups. It may also promote noted individuals as models, serving to both motivate and reward contributors.

In a virtual team, greater performance often results when leaders encourage recognition and reward members. Research confirms that for individuals, recognition not only improves motivation but also sustains greater levels of performance (Burt, 1987; Cohen and Bailey, 1997). As a consequence, teams exhibit greater interdependence and collaboration. In virtual teams, these processes often result in greater internal boundary spanning, ultimately contributing to cohesion and a shared mental model, crucial to a team's performance (Ancona and Caldwell, 1988; Yukl, 1999). When virtual team leaders promote recognition, they encourage individual and collective success.

ADVOCACY

Advocacy extends boundary spanning by exploiting internal and external communication channels. It allows leaders or other team members the chance to promote, plead, or argue in support of a subgroup or individual member's efforts. It includes aspects of buffering, impression management, and promotion of team members. Externally, advocacy attempts to secure external support for the team as well as its individual members. Together with recognition, advocacy can build an *esprit de corps*, reducing virtual distance between members.

Buffering introduces a barrier designed to relieve pressure or stress between parties. Internally, buffering mediates differences—emotional, philosophical, or physical—that may arise among members of a team. Differences may be based on individual values or actual work product, but, regardless of why conflicts emerge, team leaders or others serve as intermediaries aiming to diffuse issues coolly.

Rather than relying on one's self to promote a positive image, advocacy uses impression management to present images of others, singularly and collectively, to third parties. The purpose is the same—to create a positive image of the individual or group.

Promotion of individuals is a form of recognition designed to appeal to third parties. Promotional behaviors aimed at outside parties are used not only to enhance an image as with impression management but also to secure future benefits for individuals or the team.

PRELIMINARY FINDINGS

In a study of 266 individuals in 42 teams conducted over two years, investigators found significant support for Ambassadorial Leadership. The study concluded that all five leadership behaviors contributed significantly to team effectiveness through OCB ($p \leq 0.05$). The study also found that four behaviors—internal boundary spanning, external boundary spanning, shared leadership, and advocacy—also greatly contributed to team effectiveness using POS ($p \leq 0.05$). While there is a need for further research to confirm the findings, nonetheless the results are encouraging since they identify specific behaviors that are likely to support the success of virtual teams.

Challenges and Opportunities

As with many efforts, if properly managed, challenges posed by a virtual team may sometimes be turned into opportunities. Considered from the social network perspective, a virtual team is a collection of individuals and subgroups loosely held together by a single individual, the team leader. For example, faultlines occur because of differences in a population. However, if there were no differences, there would be a single, homogenous group, with little or no opportunity to learn since everyone in the group knows what everyone else knows.

This dilemma presents us with a conundrum. What is preferable—a loosely associated group with little in common? Or a homogenous team? How might we reconcile this problem to create an effective team? With this dilemma in mind, the first objective of Ambassadorial Leadership is to build a team into a cohesive structure. In social network theory, whenever two nodes (individuals or groups) exist without a shared or common member, we consider the vacancy to be a structural hole (Burt, 2004). When filled, structural holes create a bond between two nodes that previously did not exist. The bond is a channel across which information is exchanged. It is in the process of exchange that opportunities present themselves, a concept articulated by theory of diffusion. Diffusion can occur only when a group is bridged to an outside source that then introduces something new to the group.

The potential strength of a virtual team lies in its diversity. Recent studies have shown that by embracing diversity, rather than trying to eliminate it, teams are significantly more effective (Ely and Thomas, 2001; Derek and Kecia, 2004). It turns out that teams that learn from each other create solutions that otherwise might remain hidden. Team leaders that adopt Ambassadorial Leadership create environments that embrace diversity and the free exchange of information. These behaviors can be promoted among team members and between the team and outside entities. While the team leader may be successful in promoting these attributes within an extended team, equally, the parent organization must provide effective support for the leader's efforts. If the organization views diversity only as a challenge, they may undermine even the most zealous efforts of the Ambassadorial Team leader.

The Challenge	Ambassadorial Behavior	Actions	The Opportunity
Building a shared mental model when there may be a lack of unified core values	Internal boundary spanning	Promote relationship building between close and distant team members by encouraging sharing of personal information	*Team*: Embracing diversity provides a means to introduce multiple perspectives; builds team cohesion and trust
Maintaining individual core values while participating within the team		Educate local team members on differences in cultural values and communication	*Team Members*: Elucidating and explaining the differences to team members from other cultures builds team trust

(*continued*)

The Challenge	Ambassadorial Behavior	Actions	The Opportunity
		styles of remote members	and extends the personal social network
		Establish key relationships with members at remote locations who can serve as mentors and coaches	*Key Member*: Serve as liaison between team and remote members; can work as a cultural translator
Aligning team goals with various external entities	External boundary spanning	Establish communication channel to ensure information is freely exchanged	*Team*: External entities become partners—exchanging information and resources to secure best results
Maintaining skills within original discipline and relationship with actual business unit/functional group		Formulate strategic plan with distant members to develop liaison relationships with their close external groups	*Team*: Communication channels ensure that team goals remain viable
		Develop understanding of resources available from external groups	*Team Members*: As liaisons are better able to judge what resources might be beneficial (in both directions) and opens communications to support skill

The Challenge	Ambassadorial Behavior	Actions	The Opportunity
			retention/ development
Providing active leadership across dispersed team and diverse disciplines	Shared leadership	Create conditions for shared leadership at distant locations	*Team*: Brings the best talent to bear as dictated by the situation, promoting trust and building greater collaboration
Conveying unique requirements of the discipline or culture to a leader with a different background		Establish key relationships with members at remote locations who can serve as mentors and coaches	*Team Members*: Actively presenting unique viewpoints and representing the subgroups as viable collaborators
Motivating dispersed team members and subgroups	Recognition	Depending on the core values of the individual's culture provide open recognition of personal performance or subgroup performance	*Team*: Reinforces the self-worth of the individuals and/ or groups that are recognized
Establishing individual's value to functional group		Privately recognize the contributions made by an individual	*Team*: Provides a role model for other team members

(*continued*)

The Challenge	Ambassadorial Behavior	Actions	The Opportunity
		Privately or publicly (depending on core values) provide recognition of performance to the individual's functional group	*Team Members*: Recognizing the contributions of others may add to the group cohesion
Resolving conflicts that result from differences in core values	Advocacy	Monitoring conflicts and problems between close and distant members Serve as a mediator in cases of conflict	*Team*: Conflicts stemming from diversity provide opportunity for creative solutions
Ensuring team contributions are viewed favorably		Acknowledge team's contribution to organizational strategy externally	*Team*: Linking the team accomplishments to strategic goals elevates individual and team status and reflects favorably on contributing functional groups
Adjusting to changing demands of external entities		Monitor expectations (internal and external) and ensure alignment with reality	*Team*: Frequently checking with the stakeholders ensures that the end product will still have value

TIPS FOR VIRTUAL TEAM MEMBERS

1. *Embrace diversity. Recognize that novelty can introduce new opportunities.* Approach teaming positively. Although many of us are uncomfortable when faced with novel or unknown conditions, such situations can provide us with new opportunities to learn.
2. *Be considerate of diversity. Recognize that cultural differences may be manifested in team behavior.* When differences arise, discuss them, rather than taking exception. Open communication is critical as a replacement for the absence of face-to-face interaction.
3. *Be proactive. Recognize that the communication channel serves as your eyes, ears, and mouth.* The only way to discover if another team member has the resources that you need to complete your task or whether another team member has resources available to you to complete theirs is to communicate to the team.
4. *Don't dilute the value of communication. Recognize the value of good judgment.* Not all communication needs to be broadcast to the entire team. When responding to a broadcast message, consider whether there is a need to "reply to all," instead of just to the member who sent the message.
5. *Explore the team vision and goals. Recognize their value to the ultimate success of the team and each member.* Team vision and goals will provide guidance in times of uncertainty. In the absence of regular face-to-face interaction, vision and goals may serve to reinforce your personal motivation. Ultimately, the team's vision and goals should be adopted as one's own. With the adoption of the shared mental model, the team's success becomes our own success and reinforces your personal growth and self-actualization.

REFERENCES

Allen, N. J. and Meyer, J. P. (1996) Affective, continuance, and normative commitment to the organization: an examination of construct validity. *Journal of Vocational Behavior*, 49(3), 252.

Ancona, D. G. and Caldwell, D. F. (1988) Beyond task and maintenance defining external functions in groups. *Group & Organization Studies*, 13(4), 468.

Ancona, D. G. and Caldwell, D. F. (1992a) Bridging the boundary: external activity and performance in organizational teams. *Administrative Science Quarterly*, 37(4), 634.

Blau, P. M. (1964) *Exchange and Power in Social Life*. New York: Wiley.

Buchanan, B. (1974) Building organizational commitment: the socialization of managers in work organizations. *Administrative Science Quarterly*, 19(4), 533.

Burt, R. S. (1987) Social contagion and innovation: cohesion versus structural equivalence. *The American Journal of Sociology*, 92(6), 1287.

Burt, R. S. (1995) *Structural Holes: The Social Structure of Competition*. Cambridge, MA: Harvard University Press.

Burt, R. S. (2004) Structural holes and good ideas. *The American Journal of Sociology*, 110(2), 349.

Cappel, J. J. and Windsor, J. C. (2000) Ethical decision making: a comparison of computer-supported and face-to-face group. *Journal of Business Ethics*, 28(2), 95.

Cardona, P., Lawrence, B. S., and Bentler, P. M. (2004) The influence of social and work exchange relationships on organizational citizenship behavior. *Group & Organization Management*, 29(2), 219.

Cohen, S. G. and Bailey, D. E. (1997) What makes teams work: group effectiveness research from the shop floor to the executive suite. *Journal of Management*, 23(3), 239.

Cohen, S. G., Ledford, G. E., Jr., and Spreitzer, G. M. (1996) A predictive model of self-managing work team effectiveness. *Human Relations*, 49(5), 643.

Deluga, R. J. (1994) Supervisor trust building, leader–member exchange and organizational citizenship behaviour. *Journal of Occupational and Organizational Psychology*, 67(4), 315.

Deluga, R. J. (1998) Leader–member exchange quality and effectiveness ratings. *Group & Organization Management*, 23(2), 189.

Derek, R. A. and Kecia, M. T. (2004) Blending content and contact: the roles of diversity curriculum and campus heterogeneity in fostering diversity management competency. *Academy of Management Learning & Education*, 3(4), 380.

Eisenberger, R., Huntington, R., Hutchison, S., et al. (1986) Perceived organizational support. *Journal of Applied Psychology*, 71(3), 500.

Eisenberger, R., Fasolo, P., and Davis-LaMastro, V. (1990) Perceived organizational support and employee diligence, commitment, and innovation. *Journal of Applied Psychology*, 75(1), 51.

Ely, R. J. and Thomas, D. A. (2001) Cultural diversity at work: the effects of diversity perspectives on work group processes and outcomes. *Administrative Science Quarterly*, 46(2), 229.

Fiol, C. M. and O'Connor, E. J. (2005) Identification in face-to-face, hybrid, and pure virtual teams: untangling the contradictions. *Organization Science*, 16(1), 19.

Follett, M. P. (1927/1942) Leader and expert. In: Metcalf, H. C. and Urwick, L., editors. *Dynamic Administration: The Collected Papers of Mary Parker Follett*. New York: Harper & Brothers, p. 320.

Gould, S. (1979) An equity-exchange model of organizational involvement. Academy of Management. *The Academy of Management Review (pre-1986)*, 4(000001), 53.

Hampden-Turner, C. and Trompenaars, A. (1993) *The Seven Cultures of Capitalism: Value Systems for Creating Wealth in the United States, Japan, Germany, France, Britain, Sweden, and the Netherlands*. New York: Currency/Doubleday.

Hogg, M. A. and Terry, D. J. (2001) Social identity theory and organizational processes. In: Hogg, M. A. and Terry, D. J., editors. *Social Identity Processes in Organizational Contexts*. Philadelphia, PA: Psychology Press, p. 339.

Jones, R., Oyung, R., and Pace, L. (2005) *Working Virtually: Challenges of Virtual Teams*. Hershey, PA: Cybernetic Publishing.

Katz, D. and Kahn, R. L. (1978) *The Social Psychology of Organizations*. New York: Wiley.

Kanjorski, M. A. and Pugh, S. D. (1994) Citizenship behavior and social exchange. *Academy of Management Journal*, 37(3), 656.

Kidwell, R. E., Jr., Mossholder, K. W., and Bennett, N. (1997) Cohesiveness and organizational citizenship behavior: a multilevel analysis using work group and individuals. *Journal of Management*, 23(6), 775.

Lau, D. C. and Murnighan, J. K. (1998) Demographic diversity and faultlines: the compositional dynamics of organizational groups. *Academy of Management Review*, 23(2), 325.

Levinson, H. (1965) Reciprocation: the relationship between man and organization. *Administrative Science Quarterly*, 9(4), 370.

MacKenzie, S. B., Podsakoff, P. M., and Fetter, R. (1993) The impact of organizational citizenship behavior on evaluations of salesperson performance. *Journal of Marketing*, 57(1), 70.

MacKenzie, S. B., Podsakoff, P. M., and Rich, G. A. (2001) Transformational and transactional leadership and salesperson performance. *Academy of Marketing Science Journal*, 29(2), 115.

March, J. G. and Simon, H. A. (1958) *Organizations*. New York: Wiley.

Martins, L. L., Gilson, L. L., and Maynard, M. T. (2004) Virtual teams: what do we know and where do we go from here?. *Journal of Management*, 30(6), 805.

Meyer, J. P. and Allen, N. J. (1991) A three-component conceptualization of organizational commitment. *Human Resource Management Review*, 1(1), 61.

Meyer, J. P., Allen, N. J., and Smith, C. A. (1993) Commitment to organizations and occupations: extension and test of a three-component conceptualization. *Journal of Applied Psychology*, 78(4), 538.

Organ, D. W. (1988) *Organizational Citizenship Behavior: The Good Soldier Syndrome*. Lexington, MA: Lexington Books.

Pearce, C. L. and Sims, H. P., Jr. (2002) Vertical versus shared leadership as predictors of the effectiveness of change management teams: an examination of aversive, directive, transactional, transformational, and empowering leader behaviors. *Group Dynamics*, 6(2), 172.

Podsakoff, P. M., Ahearne, M., and MacKenzie, S. B. (1997) Organizational citizenship behavior and the quantity and quality of work group performance. *Journal of Applied Psychology*, 82(2), 262.

Schmidt, J. B., Montoya-Weiss, M. M., and Massey, A. P. (2001) New product development decision-making effectiveness: comparing individuals, face-to-face teams, and virtual teams. *Decision Sciences*, 32(4), 575.

Shore, L. M. and Shore, T. H. (1995) Perceived organizational support and organizational justice. In: Cropanzano, R. and Kacmar, K. M., editors. *Organizational Politics, Justice, and Support: Managing the Social Climate of the Workplace*. Westport, CT: Quorum Books, p. 240.

Shore, L. M. and Tetrick, L. E. (1991) A construct validity study of the survey of perceived organizational support. *Journal of Applied Psychology*, 76(5), 637.

Sobel Lojeski, K., Reilly, R., and Dominick, P. (2006) The role of virtual distance in innovation and success. *Proceedings of the 39th Annual*

Hawaii International Conference on System Sciences (HICSS '06), Kauai, HI.

Straus, S. G. and McGrath, J. E. (1994) Does the medium matter? The interaction of task type and technology on group performance and member reactions. *Journal of Applied Psychology*, 79(1), 87.

Wayne, S. J., Shore, L. M., and Linden, R. C. (1997) Perceived organizational support and leader–member exchange: a social exchange perspective. *Academy of Management Journal*, 40(1), 82.

Wayne, S. J., Tetrick, L. E., Shore, L. M., et al. (2002) The role of fair treatment and rewards in perceptions of organizational support and leader–member exchange. *Journal of Applied Psychology*, 87(3), 590–598.

Yukl, G. (1999) An evaluation of conceptual weaknesses in transformational and charismatic leadership theories. *Leadership Quarterly*, 10(2), 285.

CHAPTER

3

PEER AND SELF-ASSESSMENT

PAUL RESTA AND HAEKYUNG LEE

For instructors, evaluation of teamwork rarely poses a challenge. Faculty can easily establish criteria to assess the quality of a team's intellectual product or performance. Individual accountability, on the other hand, is not nearly as simple. It is often difficult to measure the level of contribution made by individual team members.

As noted by Johnson and Johnson (2004), to be successful, several conditions are essential for face-to-face as well as online collaborative learning teams—positive interdependence, interaction, teamwork, social skills, and individual and group accountability. Peer assessment offers a means of providing individual accountability. Self-assessment also helps learners reflect on their own performance and enhances their accomplishments. This chapter discusses major issues related to the use of peer and self-assessment in learning teams and describes an open-source, web-based assessment system designed to help an instructor develop and use peer and self-assessment instruments. It also presents key findings of a study by the authors of the perceptions of peer and self-assessment by college students in an online course. Practical suggestions are

Virtual Teamwork: Mastering the Art and Practice of Online Learning and Corporate Collaboration. Edited by Robert Ubell

offered to instructors who may wish to use peer and self-assessment in their courses.

COLLABORATIVE LEARNING

Collaborative learning has played a key role in progressive education since the early nineteenth century. According to Slavin (1997), it is perhaps one of the greatest success stories in the history of education. Since the early 1970s, the number and quality of studies on collaborative education has been increasing, and it is currently one of the principal areas of research in education. Many scholars have concluded that collaborative learning can be even more effective than traditional instruction (Rogoff, 1990; Freeman, 1995; Garvin et al., 1995; Lejk et al. 1996; Rafiq and Fullerton, 1996; Johnson and Johnson, 2004). Today, it is recognized by many as a critical strategy for preparing students with the skills to compete in our knowledge-based global society (Partnership for 21st Century Skills, 2008).

The collaborative approach to learning encourages learners to work together on tasks, promoting individual learning by engaging them in collective processes. It offers opportunities to learn using dialogue and discussion while exploring diverse ideas and experiences. Process-driven participants work together to solve problems, accomplish tasks, and create intellectual results, often not easily accomplished by a single individual. This opens the door to diverse perspectives, deepening understanding, sharpening judgment, and extending knowledge (Cowie and Rudduck, 1988). Collaborative education can yield outcomes beyond academic achievement, increasing competence in working with others and enhancing leadership skills. It engages learners to think about why they are learning and for whom they are learning (Resta et al., 2002).

Many have embraced powerful new social networking tools to facilitate collaborative learning. The trend represents a confluence of new collaborative tools (Johnson and Johnson, 2004), together with the wider acceptance of constructivist teaching and learning methods (Kirschner et al., 2004), as well as the need to create new and more engaging learning environments (Oblinger and Oblinger, 2005). New technologies allow students to share their ideas and communicate with each other, eliminating constraints of time and location (Phelps et al., 1991; Bates, 1995; Crook, 1996; Liu et al., 1999; Dede, 1996). A number of studies have examined the benefits of group collaboration, exposing students to other points of view, facilitating active learning, and forging interpersonal relationships and individual responsibilities (Jacques, 1991; Michaelsen, 1992; Mello, 1993; Harvey and Green, 1994; Freeman, 1995; Garvin et al., 1995).

PEER AND SELF-ASSESSMENT

Individual and group accountability are essential for successful virtual teams. While group assessment may be easily achieved when evaluating the quality of an intellectual product created by the team as a whole, individual accountability is far more challenging since it is often difficult to determine the actual contribution made by individual members. Peer and self-assessment may be an effective alternative to conventional faculty judgment, not only for the group but for individual members also. In peer assessment, team members apply a set of standards in order to make critical judgments about the work of the collective as well as the contributions of others in the group (Sluijsmans et al., 1999).

The literature suggests that peer assessment can improve student learning by encouraging students to consider the objectives of assessment as well as the purposes served by the course itself (Topping et al., 2000). It often confronts students with questions about what constitutes a good piece of work, opening them to feedback about their performance by other team members (Searby and Ewers, 1997). Peer assessment can take the mystery out of the process, giving students an appreciation of why grades are awarded (Brindley and Scoffield, 1998). It may provide a better understanding of what is required to achieve a particular standard and what instructors are looking for (Falchikov, 1995; Race, 1998; Hanrahan and Isaacs, 2001). It also gives students a chance to critique writing styles, techniques, ideas, and abilities, encouraging them to learn from both the exemplary performances as well as the mistakes made by others (Race, 1998). It alerts participants to dilemmas instructors face in grading (Billington, 1997; Hanrahan and Isaacs, 2001), highlighting the importance of presenting work in a clear, logical format (Brindley and Scoffield, 1998; Race, 1998), and encourages students to reflect on their own achievement (Dochy et al., 1999). When effective, peer and self-assessment can increase students' understanding and self-confidence as well as the quality of their work (Mowl and Pain, 1995; Dochy et al., 1999; Topping et al., 2000). Fostering interdependent learning, peer assessment can build collaboration, rather than competition, often generating effective interpersonal competence (Heron, 1981, p. 86).

Self-assessment gives students the chance to reflect on their own performance in the same way that they judge the work of their peers. In self-assessment, learners take responsibility for monitoring and making judgments about their own learning (Resta et al., 2002), requiring them to think critically about what they are learning, to identify standards of performance, and to apply these standards to their own work.

Teamwork Assessment Scale

Using a pool of items drawn from prior research on performance in virtual teams, Resta and DeHoyos (2002) created an online teamwork assessment scale (TAS). The result, following repeated revisions and analysis to refine and validate it, is a scale of 16 assessment items categorized in 3 dimensions—social interaction, task management, and trust.[1]

ONLINE ASSESSMENT SYSTEM

Because it can be time consuming and difficult for instructors to create their own assessment instrument, an online open source tool for self-, peer, and project assessment was created. The online assessment system (OAS) enables faculty to integrate self-, peer, and project assessment into courses populated by virtual teams (Resta, 2005). Once faculty mount it online, students, after completing collaborative tasks, projects, course modules, or lessons, can immediately enter evaluations of themselves and other members of their team. It also helps them assess the quality of projects created by their team as well as those of other groups. The tool enables instructors to create traditional or rubric-based assessment instruments, customized to the content of the course and quality criteria. Assessment results are anonymous. Only the individual student and the instructor are able to view evaluations entered by that student. Figure 3.1 provides an example of the online peer and self-assessment webpage used by students to rate themselves and the members of their virtual team.

Assessment data are automatically entered in a database and summarized. Based on average ratings by team members, results are presented to each student in a tabular and graphic form. The system shows students their overall average score from peers, average score of peer and self-assessment items

[1]Items use a 5-point scale, between the two extremes of "Never" and "Always," to reflect personal efforts and group contributions. Responses range from 1 for "Never" to 2 for "Seldom," 3 for "Sometimes," 4 for "Frequently," and 5 for "Always."

- Always demonstrates the quality, you would give a score of **5**
- Frequently demonstrates the quality, you would give a score of **4**
- Sometimes demonstrates the quality, you would give a score of **3**
- Seldom demonstrates the quality, you would give a score of **2**
- Never demonstrates the quality, you would give a score of **1**

Questions	Ava	Ava	Ava	Ava
	Ava	Jasmine	William	Linda
1. Takes active role on initiating ideas or actions.	Select	Select	Select	Select
2. Is willing to take on task responsibilities.	Select	Select	Select	Select
3. Is willing to frequently share ideas and resources.	Select	Select	Select	Select
4. Accepts responsibilities for tasks determined by the group.	Select	Select	Select	Select
5. Helps promote team esprit de corps.	Select	Select	Select	Select
6. Respects differences of opinions and backgrounds, and is willing to negotiate and make compromises.	Select	Select	Select	Select
7. Provides leadership and support whenever necessary.	Select	Select	Select	Select
8. Acknowledges good works of other members and provides positive feedback.	Select	Select	Select	Select
9. Is willing to work with others for the purpose of group success.	Select	Select	Select	Select
10. Communicates online in friendly tone.	Select	Select	Select	Select
11. Keeps in close contact with the rest of the team so that everyone knows how things are going.	Select	Select	Select	Select
12. Produces high-quality work.	Select	Select	Select	Select
13. Meets the deadlines of our team.	Select	Select	Select	Select
14. Sensitive to the needs and feelings of other members of the team.	Select	Select	Select	Select
15. Understand problems and responds with helpful comments.	Select	Select	Select	Select
16. Openly shares needs and feelings with team members.	Select	Select	Select	Select

Comments:

Jasmine :

William :

Linda :

Cancel Reset Submit Rating

FIGURE 3.1. Peer and self-assessment webpage.

(Figure 3.2), comparison graph of peer and self-assessment scores (Figure 3.3), and comments from peers.

Results help students identify areas of strength and weakness in their participation and contributions.

Because many students have had little or no prior experience in peer and self-assessment, before they enter the process, it's best to help them understand its benefits as well as common errors. For many, *leniency* and *friendship* often result in overestimating the strength of peers. For others, *severity* may generate damaging assessments. For still others, *differential interpretation* of established assessment criteria—misunderstanding what constitutes quality performance—can lead to other assessment problems.

Your Results

Average per Question

No.	Question	Peer Average
1	Takes active role on initiating ideas or actions.	4.67 / 5
2	Is willing to take on task responsibilities.	4.67 / 5
3	Is willing to frequently share ideas and resources.	4.67 / 5
4	Accepts responsibilities for tasks determined by the group.	4.67 / 5
5	Helps promote team esprit de corps.	4 / 5
6	Respects differences of opinions and backgrounds, and is willing to negotiate and make compromises.	4.33 / 5
7	Provides leadership and support whenever necessary.	4.67 / 5
8	Acknowledges good works of other members and provides positive feedback.	4.67 / 5
9	Is willing to work with others for the purpose of group success.	4.67 / 5
10	Communicates online in friendly tone.	4.67 / 5
11	Keeps in close contact with the rest of the team so that everyone knows how things are going.	4.33 / 5
12	Produces high-quality work.	5 / 5
13	Meets the deadlines of our team.	5 / 5
14	Sensitive to the needs and feelings of other members of the team.	5 / 5
15	Understand problems and responds with helpful comments.	5 / 5
16	Openly shares needs and feelings with team members.	5 / 5

FIGURE 3.2. Average score of peer and self-assessment items.

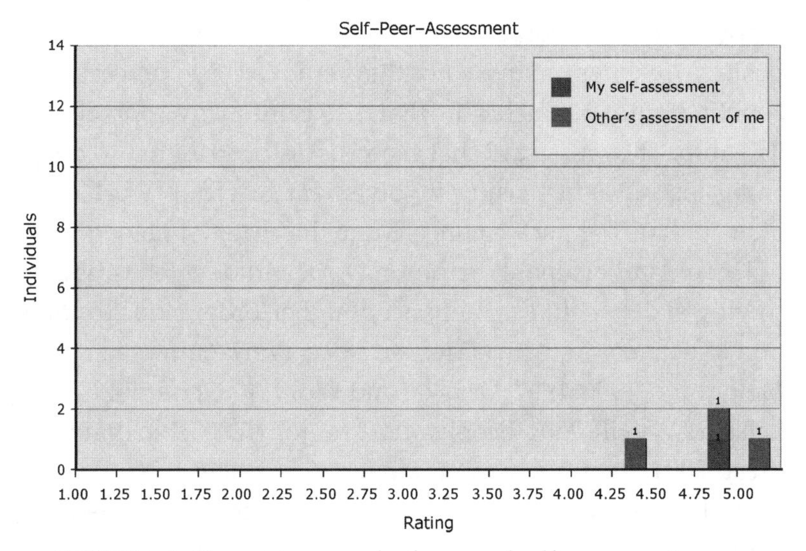

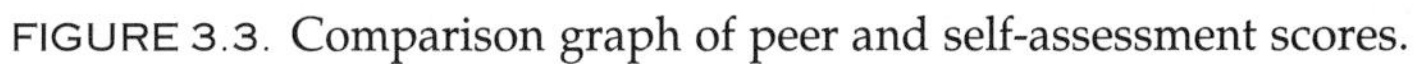

FIGURE 3.3. Comparison graph of peer and self-assessment scores.

It's also important to provide students with an opportunity to practice peer assessment using established protocols. It's wise to create a hypothetical team member and allow students to judge his or her fictional performance. It's helpful if the mock scenario is closely related to the activity of the team. Members are then asked to assess the hypothetical performance using a peer assessment instrument. Team members are then asked to discuss their evaluation and to try to arrive at a consensus. This exercise helps build confidence, as well as positive attitudes toward the process. Students without prior experience in peer assessment often feel uncomfortable without such training.

STUDENT PERCEPTION

A number of recent studies have examined student perception and levels of satisfaction with peer and self-assessment in online learning. Results reveal a mixed picture. While some indicated that they felt positive about the experience, others reported a less favorable view. Among those who expressed positive feelings, many said they enjoyed the process and benefited from it, concluding that they came away appreciating its value. They also said that the process contributed to deepening learning, increasing involvement in group work, and, often, enhancing performance. Many indicated it helped them reflect on and evaluate their own work and develop problem-solving skills, including enhanced higher-order skills to function effectively in teams (Stefani, 1994; Warkentin et al., 1995; Melograno, 1996, 1997; Druskat and Wolff, 1999; Gatfield, 1999; Lejk and Wyvill, 2001; Brooks and Ammons, 2003; Bloxham and West, 2004; Evans et al., 2005; Li and Steckelberg, 2006).

Others, however, revealed serious concerns and negative reactions. Some students felt they lacked the ability to judge the work of their peers, lacking expert knowledge of the content, and concluding that they did not posses enough experience to be objective. Some claimed that personal bias, peer pressure, friendships, and relationships with others limited their capacity to judge fairly. Many said that because of limited assessment training, as well as complex and unclear criteria, their evaluations may have been misleading. Others claimed that the process was too time consuming (Cheng and Warren, 1997; Brindley and Scoffield, 1998; Lopez-Real and Chan, 1999; Daniels and Magarey, 2000; Hanrahan and Isaacs, 2001; Rees et al., 2002; Sluijsmans et al., 2002).

In an effort to determine the strength of findings reported in the literature, the authors introduced self- and peer assessment in a graduate-level online course, "Computer Supported Collaborative Learning," in which 14 students from diverse ethnic, gender, and institutional populations, as well as learning and computer skill backgrounds, participated. Based on student background and computer skills, the instructor divided the class into six teams of three to four members each.

The course was collaborative, exploiting online communication tools. At the end of each project, students conducted online peer and self-assessment. Assessments were performed anonymously using the OAS described earlier.

Primary data was drawn from in-depth, semi-structured face-to-face and online videoconferencing interviews. Researchers asked participants to describe their experiences and views of the process. In addition, responses were collected from written reflections and portfolios, messages posted to an online discussion board, peer comments shown in a web-based assessment tool, and peer and self-assessment scores.[2] Figure 3.4 summarizes principal student perceptions on the use of peer and self-assessment in the course.

[2]The study mainly used techniques and procedures based on grounded theory, originally developed by Barney Glaser and Anselm Strauss (Strauss and Corbin, 1998). In grounded theory analysis, the process of generating theory from data is delimited by a set of rigorous analytic procedures: open coding, axial coding, and selective coding. As a result of the analysis, most prominent factors related to the participants' perceptions emerged.

Factors Influencing Student Perceptions of the Peer and Self-Assessment

Learning Context	Individual Differences	Online Learning Community
Course elements Assessment practice Repeated assessment process Assessment as an assignment **Online assessment system** Anonymous assessment Computerized online assessment Assessment rubric Problems with the assessment system **Types of assessment feedback** Numerical results Textual comments **Graduate school environment** Academic context Equality with peers	**Stringency-leniency in ratings** **Objectivity of ratings** **Previous assessment experience** **Purpose of assessments** Encouraging peers Reinforcing peers strengths **Degree of self-confidence in assessing their own contributions to the group activity**	**Group composition** Diversity of group members Small group size **Engagement of group members** **Sense of community** Familiarity with group members Conscious of other group members Consideration of team building

Use of Peer and Self-Assessment

Impact of the Use of Peer and Self-Assessment on Group Collaboration

Understanding others' perspectives	Interpersonal skills for collaboration	Personal criteria for assessments
Reflections on themselves	Accountability	Level of confidence with peer and self assessment
Awareness of the assessment	Participation	Group collaboration

FIGURE 3.4. Principal student perceptions on the use of peer and self-assessment.

LEARNING CONTEXT

Results indicated that student reactions to self- and peer assessment were strongly influenced by the learning context. When students were given a chance to practice peer assessment and were able to discuss various interpretations and ratings with team members before conducting their first formal assessment, they felt comfortable with the process. They also responded well to consistency and the routine nature of the assessment procedure. Rater anonymity and the easy-to use format of the OAS also contributed to positive perceptions of peer and self-assessment.

On the whole, students responded favorably to the tool, with some reporting that the graphical display enabled them to easily access their own and others' performance (see Figure 3.2). Others preferred numerical data, while still others felt that textual comments, compared with graphical representation, were preferable (see Figure 3.3).

INDIVIDUAL DIFFERENCES

Students varied widely on how stringent or lenient they ranked others. Some indicated that they were more lenient with others, while at the same time they were more stringent with themselves. Students who had performed peer and self-assessment earlier felt more comfortable and confident with judging others, while those without prior experience said they had some difficulty in evaluating team members. Some used the tool to encourage and motivate, while a few reacted poorly to members who failed to contribute, effectively using the tool to punish them. Those who expressed a high degree of confidence in their own contributions tended to feel comfortable conducting their own self-assessment, not hesitating to give high ratings to themselves. Those who expressed low self-confidence about their performance had difficulty assessing themselves. Participants indicated they had little or no difficulty in performing peer assessment for high-performing members, but felt uncomfortable giving ratings suggesting improvement to those who worked hard but did not come up to the mark.

Online Learning Community

Students indicated that their feelings and perceptions in conducting the peer and self-assessment were influenced by group members who had different perspectives and backgrounds. The size of the team was also a variable that affected students' comfort level in completing the assessments. For instance, the students from the smaller teams indicated they were reluctant to make assessments that might make the other team members uncomfortable.

Students also indicated that the level of the engagement of group members was a factor influencing their peer evaluations. The level of comfort in conducting peer assessment was also related to the sense of community within the team. Interestingly, the sense of community for some students contributed to their feeling more comfortable in providing honest feedback to their team members, while for others, the strong sense of community caused them to be more concerned about how their feedback might affect group members' feelings, team building, and group interaction.

Impact on Teams

The study sought to understand how peer assessment may have had an impact on group collaboration. Many said that it helped them *understand how other perspectives* related to their individual contributions, claiming that peer assessment gave them an opportunity to learn how other members perceived their contributions. They said that it encouraged them to collaborate more actively and helped them focus on areas requiring improvement. They indicated that it increased their awareness that their participation and contributions were being observed by others and assessed by other team members, increasing their sense of accountability. Over the semester, it helped to increase their confidence in requesting, giving, and receiving feedback. In the end, they claimed that it strengthened group collaboration by enhancing relationships, fostering group cohesion, and helping them to work together to support collaborative learning. For instructors, awareness of student attitudes about peer and self-assessment may help them use these assessments more effectively in online classes.

SUGGESTIONS FOR FUTURE EDUCATIONAL PRACTICE

TIP 1: GROUP SIZE

Many studies conclude that peer assessment is useful and highly effective in small groups (Ferris and Hess, 1985; Melograno, 1996; Persons, 1998; Lopez-Real and Chan, 1999). Yet in our study, small groups showed stress when students conducted peer assessments. As members of small groups, some said they could easily identify who was responsible for each rating, even though the online system ostensibly protected raters with anonymity. As a result, they tended to rate peers more generously, causing scores to be inflated. Given these concerns, instructors should be alert to group size when forming teams, especially if they intend to introduce peer and self-assessment. Should small groups be formed, faculty will need to carefully monitor performance of individual members, rather than rely solely on peer ratings, and take care in interpreting assessment results.

TIP 2: ACCURACY AND BIAS

Some students in the study were concerned that in rating other team members, they relied too heavily on their own judgment, concluding that it was difficult for them to be objective. Others worried that since the results of peer and self-assessment accounted for a significant part of their final grades, this might lead to a tendency to judge themselves and others dishonestly. Students also felt that personal relationships and alternative interpretations of assessment criteria might bias their evaluations.

One strategy to minimize bias effects is to provide students with training on peer assessment before conducting the actual assessments, unless the instructor is confident that the students have had prior experience with the peer assessment process. Bias effects may also be minimized by developing assessment criteria that are clear and well understood by the students before conducting the assessment and also by requiring the student evaluators to cite specific evidence of student work supporting their assessments. The instructor may also reduce bias effects by imposing consequences for students who do not follow the criteria in conducting their assessments.

TIP 3: ANONYMITY

For effective peer assessment, students insist on strict anonymity. These findings are consistent with the work of Davies (2002), who found that lack of anonymity contributes directly to negative feelings about peer assessment. Students are uncomfortable with rating or critiquing their peers or may feel obligated to assign friends high scores unless anonymity is strictly preserved. Based on the literature and the students' perceptions, it is clear that instructors need to consider anonymity a critical element in the use of peer assessment.

TIP 4: FORMATIVE ASSESSMENT

Most students valued peer and self-assessment as a kind of formative assessment, enabling them to understand what they were doing well and identify areas requiring improvement. As noted by Boud (1990), encouragement and feedback from peers give students a chance to learn more effectively. The use of peer and self-assessment at multiple points during virtual team projects helps participants reflect on their own learning process and enables them to identify areas for improvement and to become more involved in the group.

Peer and self-assessment, as a formative assessment, can also challenge troublesome free riders (Brooks and Ammons, 2003). Should assessment results show that some team members are hitchhiking while others are doing most of the work, instructors might meet with teams or individual members to explore better ways to distribute the workload (Ohland et al., 2005).

SUMMARY

In recent years, there has been an explosion in the number of online and blended courses offered by institutions of higher education. There has also been a growing recognition that such courses need to be more interactive and include student collaboration as important course elements. One of the major challenges to instructors using online collaborative learning is that of assessing the contributions of individual members within a virtual learning team. Peer and self-assessment, when appropriately used, may serve as an effective means of providing students with useful information about their performance and increasing their accountability to the work of the team. It also provides an important source of information, enabling instructors to better understand the individual contributions of each team member. This chapter discusses the variables related to effective performance of virtual teams and ways to assess them. It describes an open-source, web-based assessment system designed to facilitate the development and implementation of peer and self-assessment instruments and presents results of a study of the perceptions of higher education students related to peer and self-assessment. Suggestions are offered to instructors who may be considering the use of peer and self-assessment in their courses.

REFERENCES

Bates, A. W. (1995) *Technology, Open Learning and Distance Education*. London: Routledge.

Billington, H. L. (1997) Poster presentations and peer assessment: novel forms of evaluation and assessment. *Journal of Biological Education*, 31(3), 218–220.

Bloxham, S. and West, A. (2004) Understanding the rules of the game: marking peer assessment as a medium for developing students' conceptions of assessment. *Assessment & Evaluation in Higher Education*, 29(6), 721–733.

Boud, D. J. (1990) Assessment and the promotion of academic values. *Studies in Higher Education*, 15(1), 101–111.

Brindley, C. and Scoffield, S. (1998) Peer assessment in undergraduate programs. *Teaching in Higher Education*, 3(1), 79–89.

Brooks, C. M. and Ammons, J. L. (2003) Free riding in group projects and the effects of timing, frequency, and specificity of criteria in peer assessments. *Journal of Education for Business*, 78(5), 268–263.

Cheng, W. and Warren, M. (1997) Having second thoughts: student perceptions before and after a peer assessment exercise. *Studies in Higher Education*, 22, 233–239.

Cowie, H. and Rudduck, J. (1988) *Cooperative Group Work: An Overview*. London: BP Educational Services.

Crook, C. (1996) *Computers and the Collaborative Experience of Learning*. London: Routledge.

Daniels, L. A. and Magarey, A. (2000) The educational and vocational role of peer assessment in the training and professional practice of dietitians. *Australian Journal of Nutrition & Dietetics*, 57(1), 18–22.

Davies, P. (2002) Using student reflective self-assessment for awarding degree classifications. *Innovations in Education and Teaching International*, 39(4), 307–319.

Dede, C. (1996) Emerging Technologies in Distance Education for Business. *Journal of Education for Business*, 71(4), 197–204.

Dochy, F., Segers, M., and Sluijsmans, D. (1999) The use of self-, peer and co-assessment in higher education: a review. *Studies in Higher Education*, 24(3), 331–350.

Druskat, V. U. and Wolff, S. B. (1999) Effects and timing of developmental peer appraisals in self-managing work groups. *Journal of Applied Psychology*, 84(1), 58–74.

Evans, A. W., McKenna, C., and Oliver, M. (2005) Trainees' perspectives on the assessment and self-assessment of surgical skills. *Assessment & Evaluation in Higher Education*, 30(2), 163–174.

Falchikov, N. (1995) Peer feedback marking: developing peer assessment. *Innovations in Education and Training International*, 32, 175–187.

Ferris, W. P. and Hess, P. W. (1985) Peer evaluation of student interaction in organizational behavior and other courses. *The Organizational Behavior Teaching Review*, 9, 74–82.

Freeman, M. (1995) Peer assessment by groups of group work. *Assessment and Evaluation in Higher Education*, 20(3), 289–300.

Garvin, J. W., Butcher, A. C., Stefani, L. A. J., et al. (1995) Group projects for first-year university students: an evaluation. *Assessment & Evaluation in Higher Education*, 20, 273–288.

Gatfield, T. (1999) Examining Student satisfaction with group projects and peer assessment. *Assessment and Evaluation in Higher Education*, 24(4), 365–377.

Hanrahan, S. J. and Isaacs, G. (2001) Assessing self- and peer-assessment: the students' views. *Higher Education Research & Development*, 20(1), 53–70.

Harvey, L. and Green, D. (1994) *Employer satisfaction*. Birmingham: University of Central England.

Heron, J. (1981) Assessment revisited. In: Boud, D. J., editor. *Developing Student Autonomy in Learning*. London: Kogan Page.

Jacques, D. (1991) *Learning in Groups* (2nd ed.). London: Kogan Page.

Johnson, D. W. and Johnson, R. T. (2004) Cooperation and the use of technology. In: Jonassen, D.editor. *Handbook of Research for Educational Communications and Technology* (*2nd ed.*, pp. 785–811), Mahwah, NJ: Lawrence Erlbaum Associates.

Kirschner, P. A., Martens, R. L., and Strijbos, J. W. (2004) CSCL in higher education? A framework for designing multiple collaborative environments. In: Strijbos, J. W. Kirschner, P. A. and Martens, R. L., editors. *What We Know about CSCL: And Implementing It in Higher Education* (pp. 3–30). Boston, MA: Kluwer Academic Publishers.

Lejk, M. and Wyvill, M. (2001) The effect of the inclusion of self assessment with peer assessment of contributions to a group project: a quantitative study of secret and agreed assessments. *Assessment & Evaluation in Higher Education*, 26(6), 551–561.

Lejk, M., Wyvill, M., and Farrow, S. (1996) A survey of methods of deriving individual grades from group assessments. *Assessment & Evaluation in Higher Education*, 21, 267–280.

Li, L. and Steckelberg, A. L. (2006) Perceptions of web-mediated peer assessment. *Academic Exchange Quarterly*, 10(2), 265–270.

Liu, E. Z. F., Chiu, C. H., Lin, S. S. J., and Yuan, S. M. (1999) Student participation in computer science courses via the Networked Peer Assessment System (NetPeas). *Proceedings of the ICCE'99* (Vol. 1, pp. 774–777), Amsterdam: IOS Press.

Lopez-Real, F. and Chan, Y.-P.R. (1999) Peer assessment of a group project in a primary mathematics education course. *Assessment & Evaluation in Higher Education*, 24(1), 67–79.

Mello, J. A. (1993) Improving individual member accountability in small group settings. *Journal of Management Education*, 17, 253–259.

Melograno, V. J. (1996) *Designing the physical education curriculum* (3rd ed.). Champaign, IL: Human Kinetics.

Melograno, V. J. (1997) Integrating assessment into physical education teaching. *Journal of Physical Education, Recreation and Dance*, 68, 34–37.

Michaelsen, L. K. (1992) Team learning: a comprehensive approach for harnessing the power of small groups in higher education. *To Improve the Academy*, 11 107–122.

Mowl, G. and Pain, R. (1995) Using self and peer assessment to improve students' essay writing: a case study from geography. *Innovation in Education and Training International*, 32(4), 324–335.

Oblinger, D. G. and Oblinger, J. L., editors, (2005) *Educating the Net generation*. Washington, DC: Educause.

Ohland, M. W., Layton, R. A., Loughry, M. L., and Yuhasz, A. G. (2005) Effects of behavioral anchors on peer evaluation reliability. *Journal of Engineering Education*, 94(3), 319–325.

Partnership for 21st Century Skills (2008) *21st Century Skills Education and Competitiveness: A Resource and Policy Guide*. Tucson, AZ: Partnership for 21st Century Skills.

Persons, O. S. (1998) Factors influencing students' peer evaluation in cooperative learning. *Journal of Education for Business*, 73, 225–229.

Phelps, R. H., Wells, R. A., Ashworth, R. L., and Hahn, H. A. (1991) Effectiveness and costs of distance education: using computer-mediated communication. *American journal of Distance Education*, 5(3), 7–19.

Race, P. (1998) Practical pointers on peer-assessment. In: Brown, S., editor. *Peer Assessment in Practice* (pp. 102). Birmingham, SEDA.

Rafiq, Y. and Fullerton, H. (1996) Peer assessment of group projects in civil engineering. *Assessment and Evaluation in Higher Education*, 21(1), 69–81.

Rees, C., Sheard, C., and McPherson, A. (2002) Communication skills assessment: the perceptions of medical students at the University of Nottingham. *Medical Education*, 36(9), 868–878.

Resta, P. (2005) Development and validation of a web-based tool for individual and group accountability. *Proceedings of American Educational Research Association 2005*, San Francisco, USA.

Resta, P., Awalt, C., and DeHoyos, M. (2002) Self and peer assessment in an online collaborative learning environment. *Proceedings of World Conference on E-Learning in Corporate, Government, Healthcare, and Higher Education 2002* (pp. 682–689). Norfolk, VA: AACE.

Rogoff, B. (1990) *Apprenticeship in Thinking: Cognitive Development in Social Context*. New York: Oxford University Press.

Searby, M. and Ewers, T. (1997) An evaluation of the use of peer assessment in higher education: a case study in the School of Music, Kingston University. *Assessment and Evaluation in Higher Education*, 22(4), 371–383.

Slavin, R. E. (1997) *Educational Psychology: Theory and Practice* (5th ed.). Needham Heights, MA: Allyn and Bacon.

Sluijsmans, D., Brand-Gruwel, S., and van Merriënboer, J. J. G. (2002) Peer assessment training in teacher education: effects on performance and perceptions. *Assessment & Evaluation in Higher Education*, 27(5), 443–454.

Sluijsmans, D., Dochy, F., and Moerkerke, G. (1999) Creating a learning environment by using self-, peer- and co-assessment. *Learning Environments Research*, 1, 293–319.

Stefani, A. J. (1994) Peer, self and tutor assessment: relative reliabilities. *Assessment and Evaluation in Higher Education*, 19(1), 69–75.

Strauss, A. and Corbin, J. (1998) *Basic of Qualitative Research: Techniques and Procedures for Developing Grounded Theory* (2nd ed.). Newbury Park, CA: Sage.

Topping, K. J., Smith, E. F., Swanson, I., and Elliot, A. (2000) Formative peer assessment of academic writing between postgraduate students. *Assessment and Evaluation in Higher Education*, 25(2), 146–169.

Warkentin, R. W., Griffin, M. M., Quinn, G. P., and Griffin, B. W. (1995) *An exploration of the effects of cooperative assessment on student knowledge structure*. Paper presented at the Annual Meeting of the American Educational Research Association, San Francisco, CA.

CHAPTER 4

MITIGATING CONFLICT

RICHARD DOOL

My cell phone rings at 10:37 p.m. I answer reluctantly, anticipating the reason for the call. Team assignments are due tonight, and a call this late is seldom made to tell me everything is wonderful. Sure enough, there is trouble in "teamland" and panic is setting in. A teammate has gone AWOL, and her part is not done. Professors using team assignments in online courses often get calls or e-mails late in the game.

I have had the privilege—and at times, the agony—of overseeing more than 250 graded virtual team assignments. After enduring more than 20 instances of conflicts in student teams and occasionally making mistakes in handling them, I concluded that I needed to do something to reduce or manage conflict. Calling upon the research literature, entering discussion with my peers, and through experimentation, I have learned how to mitigate much of the inevitable conflict.

In a poll of more than 300 students at 4 universities, I asked about their experience with team assignments. Participants were offered three choices—positive, mixed, or negative. The results (Table 4.1) showed that, while 37% reacted positively, 63% revealed either a mixed or negative experience. The

Virtual Teamwork: Mastering the Art and Practice of Online Learning and Corporate Collaboration. Edited by Robert Ubell

TABLE 4.1. STUDENT ATTITUDES TOWARD VIRTUAL TEAM ASSIGNMENTS

Students	Participated in virtual team assignments	Enjoyed it and said experience was positive	Mixed reaction: worthwhile with some negative feelings	Disliked it and said experience was negative
303	252 (83%)	93 (37%)	51 (20%)	108 (43%)

primary reason students offer is difficulty in getting everyone on the same page, unclear instructions and expectations, and the uncomfortable fact that their grade depends on others (see Chapter 3).

Given potential conflicts and poor student reaction, why do instructors persist on giving team assignments? While some faculty believe in the value of collaborative education as an effective practice, encouraging students to perform as positive team members, others say it also helps reduce instructor's grading load. Because students are destined to participate increasingly in virtual teams in industry (see Chapter 10), it's wise to embed basic teaming skills in online classes.

Reducing conflict may encourage positive reactions that will serve as a basis for future growth in teaming skills. Virtual teams in organizations of all sizes and orientation have grown significantly in recent years. Almost 70% of my online students have reported working in teams in some form in the last 12 months. As globalization compresses time and space, the use of virtual teams continues to rise. While I believe that learning is improved by being an effective member of a team, especially in a virtual team, it has also become a necessary career competency.

But virtual teams can also be challenging, with obstacles related to trust, communication, dependence on technology, time management, and team cohesiveness (Smith, 2008). Building trust virtually is far more difficult than face to face owing to the lack of nonverbal cues, including proximity, consistency, and observation (see Chapters 2 and 7). In virtual teams, holding incidental meetings is rare as are ad hoc social exchanges, typical in face-to-face settings that can form trusting relationships. It is equally difficult to detect individual expectations in virtual teams. Expectations are often implicit and may not be easy to appreciate online. You may suspect mismatches when you see your team member in a classroom (Bosch-Sijtsema, 2007), but when they are at a distance, team members often have more difficulty creating a shared context, helping to shape expectations and build trust (Hinds and Bailey, 2003).

Online, communication is mediated by computers (see Chapter 7), introducing challenges, especially obstacles related to technology—software, access, response times, and poor training. Barriers in virtual teams may also come from differences in language, quality of exchange, and misunderstandings that arise from cultural, temporal, or spatial dispersion. Mortensen and Hinds (2001) note, "Individuals communicating through computer technology find it more difficult to reach consensus, are more self-absorbed, more uninhibited and pay less attention to social norms." What's more, differences in time zones, personal schedules, other commitments, inconsistent time-management capabilities, and even cultural views of the nature of time can all create conflict.

Because of these and other factors, virtual team members are more likely to experience task, role, or responsibility ambiguity (Shin, 2005). Mortensen and Hinds (2001) also note that virtual teams experience greater task conflict because of logistical difficulties and less evenly distributed information. These obstacles frequently lead to more conflict as members conclude that the difficulties they experience are caused intentionally by their peers.

SOURCES OF CONFLICT

Conflicts online may also mirror those in face-to-face teams. Most can be traced to differences in expected outcomes or grades, deliverables, roles, style, values, and resources, including available time, or personality. Because communication is often asynchronous, there seems to be more opportunity for miscommunication, much like those found in the workplace with e-mail and instant messaging. The principal difference in online student teams, as compared with those on campus, is virtuality. It significantly undermines real-time intervention or management. Faculty cannot "see" the conflict holistically, without witnessing nonverbal interactions.

Over 3 years, I followed 252 virtual team assignments, tracking all instances of evident conflict. During the period, I assigned 127 team projects, without first introducing conflict mitigation guidelines. Seventy-eight (or 61%) showed evident instances of conflict. "Evident" is a clear instance of conflict that emerges from a student complaint or observed in a group forum.

Team members "going silent" is the principal source of conflict (42%). The second most frequent was perceived lack of quality of work performed by others (31%) (see Table 4.2). Unfortunately, the third most prominent complaint is accusations of plagiarism (12%).

STUDENT REACTIONS TO TEAM ASSIGNMENTS

The beauty of online education is the opportunity to explore deeper and richer exchanges among students. Online instructors know that the level of interaction and the depth of exchange often exceed that of on-campus courses.

> The depth and frequency of intellectual exchange with your classmates exceeds that of traditional graduate school experiences. My learning

TABLE 4.2. SOURCES OF CONFLICT IN VIRTUAL TEAM ASSIGNMENTS

Tracked Virtual Team Assignments	Evident Conflicts	Sources of Conflicts
127	78 (61%)	Differences in expected outcomes and commitment (e.g., "going silent") 33 (42%)
		Differences in "quality" (deliverables) 25 (31%)
		Differences in "values" (plagiarism) 9 (12%)
		Personality conflicts 8 (10%)
		Others (including miscommunication) 3 (4%)

> team colleagues brought a wealth of experience from the private and public sector to our online discussions—which were much more lively than the traditional lecture-format learning environment.
>
> Online Student (2006)

> I have done both traditional and online programs and I found the communication and sharing of ideas to be greater and more effective in this online program.
>
> Online Student (2007)

Given these and other positive responses to the online learning experience, faculty must introduce ways to mitigate mixed or negative reactions to virtual teaming so that it does not undermine online education itself.

Recognizing that more than 60% of students surveyed in my study either disliked virtual team assignments or had a mixed reactions, it was not surprising that when students learn that their online class will include group projects, many react negatively. Some send e-mail messages saying that, while they will try to do their best, they have had bad experiences that color their response. Others even request that particular students not be assigned to their team. Some examples are as follows:

> Dr. Dool—I have to say that I do not really like group assignments. I have had some bad experiences where I have had to do most of the work. This caused me a lot of stress due to requirements of my job. I hope you will keep this in mind as you judge the assignments. I will do my best.
>
> Student X (2007)

> Professor, please do not assign me to a team with Student X, he was on my team in my last course and is full of excuses, did not do his part and hurt our team. Thank you for your consideration.
>
> Student X (2008)

> Professor, do we really have to do a team assignment? Is there an alternative? I work nights and can not really spend time in team assignments, I prefer to work alone. What can we do? I hope you will consider an alternative.
>
> Student X (2007)

When introducing team assignments, instructors are likely to experience an array of student behavior. I have witnessed the emergence of several "roles" that students assume.

Role Players

As Edward Volchok (2006) noted (see Chapter 1), students play a variety of roles in teams, some assigned and others seized without election or appointment. Volchok identified "Mussolinis, Shrinking Violets, and Rambos." While I have witnessed these as well, conflict seems to bring out a number of other personalities. I have run into the "Martyr," "Excuse-meister," "Silent Partner," and "Breathless in ... [fill in student's town here]."

The "Martyr" quickly points out that she has had to do much more than everyone else on the team because no one else seems to be taking part seriously. Her work schedule is unique and her commitments unusual. Despite that, on her own, she jumped in and assumed more responsibility for the good of the team.

The "Excuse-meister" has a great deal of creative energy, but unfortunately, he tends to focus it on why he could not do his full part. He has an array of excuses to persuade his teammates to carry more of the load, somehow rationalizing that they have less to do than he has. His excuses almost always seem to be about a sudden illness, computer challenges, or last-minute work assignments. He can be very disruptive because his excuses are often ill timed, frequently undermining team deliverables.

"Breathless in . . ." will call at the first sign of a disruptive issue and tends to cry wolf. These students can often serve as an early warning system for the instructor. They tend to overreact to deviations or lack of response from teammates, especially if they have had a prior negative team experience. They can cause an escalation of conflict by overreacting early.

The most problematic is the "Silent Partner," who is not really a partner in any meaningful sense, other than that she expects to receive the same credit as other team members, even though she has been absent from much of the work. She prefers to let others carry the load and then appears at the end with her tale of woe. "Going silent" is a team member's worst transgression, a significant source of stress and frustration, and clearly the most disruptive cause of team conflict. While past experience shows that "going silent" is a flagrant disregard of collaboration, most teams fail to address it early or appropriately enough and often fail to have a plan to deal with it other than last-minute heroics by others.

Mitigating Conflict in Student Teams

Evidence suggests that team-building exercises (Kaiser et al., 2000), establishing shared norms (Sarker et al., 2001;

Suchan and Hayzak, 2001), and specifying clear team structure (Kaiser et al., 2000) contribute to virtual team success. These elements, including clear expectations about positive team behavior and unambiguous standards for team performance—a meaningful portion of the final course grade assigned to successful teaming—are the basis for my conflict mitigation guidelines. It is suggested that the guidelines be presented to students during an orientation session.

SETTING THE STAGE

The foundation for mitigating team conflict is set before the team even meets—when the instructor creates a collaborative learning atmosphere, encouraging team members to share experiences and pool resources (Smith, 2008).

THE ROLE OF THE INSTRUCTOR

Faculty set the stage for meaningful collaboration by displaying a serious attitude toward team assignments. Students learn to perform to expectations if they are properly and consistently reminded of them. Although life gets in the way sometimes, students must adopt a "we-vs.-me" attitude and a serious commitment to the team.

The instructor must be a facilitator, boundary setter, traffic cop, and chief cheerleader. The principal role of facilitation is well understood by high-quality online instructors. If the instructor is only minimally engaged, discussion often loses energy or structure with team assignments going awry and antecedents to conflict missed. In some cases, students may be engaged and active, despite the instructor's absence, but assignments and team relationships can quickly turn sour without the presence of the instructor. Instructors should enter their online classes every day—at worst, with no more than two day's absence—and should appear visibly in team areas regularly.

Since team projects can easily get off topic or take unproductive tangents, instructors need to manage assignment boundaries by nudging team interactions on track without undermining student engagement.

Playing traffic cop and chief cheerleader are two acts in the same drama. Instructors must regulate traffic flow, team expectations, perspectives, and activity. If the team is slow to start, instructors must remind students to engage. If the quantity, quality, or visible interaction is drifting, instructors need to energize team members. If individual students are not doing their part, instructors must alert them to get them to get on track. Cheerleading is just as important. Creating a positive environment is critical to managing conflict. Praise is best delivered to members of the team publicly, while criticism is best given in private. To motivate teams, it's good to acknowledge focus and commitment frequently.

Faculty can take certain preventative actions to set a team's productivity in motion. The most important is expressing the purpose and expectations of team assignments in a well-thought-out syllabus, with an explicit guide to team expectations. A detailed syllabus facilitates knowledge exchange between members as well as mutual understanding of requirements (Hron et al., 2000).

In addition to an explicit syllabus, I post a document entitled, "There is no *'I'* in Team," stressing the professional and personal importance of being a productive and positive team member, outlining common pitfalls, examples of positive and unproductive teaming behavior, and a reminder about the we-are-all-it-together grading policy.

The grade you give students for team assignments should be significant enough to attract attention (see Chapter 3 for peer and self-assessment options). I recommend somewhere between 15% and 30% of a student's final grade. It's wise to state clearly that there is only one grade for each team assignment for all members of the team.

Because of the possibility of plagiarism, I post all the usual admonitions about academic integrity and require students to use Turnitin, a service available at many schools that matches student texts with published documents. While most faculty warn about plagiarism, it has not deterred a few, so I expect the team to monitor itself as well.

In my online classes, each student is required to send me an e-mail or post a reply to my Read This First section on my course site indicating that they understand the syllabus, teaming document, and expectations about plagiarism. Some schools use honor statements or contracts that ask students to attest to their understanding of class requirements and expectations about plagiarism. At the Stevens Institute of Technology, students are expected to append this signed statement to all assignments before submitting them:

> I pledge on my honor that I have not given or received any unauthorized assistance on this assignment/examination. I further pledge that I have not copied any material from a book, article, the Internet or any other source except where I have expressly cited the source.
>
> Signature____ Date ____

Figure 4.1 is an example of a learning contract from Seton Hall University.

In addition to my role as an educator, I strongly believe that my responsibility is also to act as an excuse eliminator, posting reminders about the importance of positive teaming before and during team activity.

I also recommend the introduction of a team charter with a portion of the grade attached to its presentation (5%) (Figure 4.2). The team jointly generates the charter based on a template I provide. It covers role assignment, skills inventory, contact and meeting information, and ways to manage conflict. Completing the team charter gives participants a chance to

Course Requirements	Yes, I have read and understand the course documents
Review the "Read Me First" documents in the Introduction Conference	☐ Yes I have
Review the course syllabus in course documents and print it out for easy reference.	☐ Yes I have
Review the assignments in the Assignments Conference.	☐ Yes I have
Review the teaming requirements in the Course Documents (There is no "I" in Team).	☐ Yes I have
Review the participation requirements in the Course Documents ("Participation and Discussion")	☐ Yes I have
Review the class site and familiarize yourself with the various sections and conferences	☐ Yes I have
Review the class schedule to ensure you understand the due dates for assignments and discussion participation.	☐ Yes I have

I have reviewed the course site and documents and understand what is required:

______________________________ Date________________
Your Signature

FIGURE 4.1. Learning contract, Seton Hall University (2008).

discuss common sources of conflict and create plans to deal with them, especially how to respond to team members who "go silent." Minimally, the charter addresses team requirements with the added benefit of pledging a commitment to its specifications. I recommend that team members complete the charter as their first assignment. It can also serve as an agenda during orientation.

With each deliverable, members are expected to turn in a team log jointly (Figure 4.3). The log documents team activities, outlining who did what. It also serves as essential documentation for two critical reasons. First, it identifies and documents for the instructor teaming disputes that may have taken place. It also allows members to explore difficulties the team may be encountering. For the final grade, the log documents disparities in student participation. The log can also help adjudicate grade appeals and illuminate the need for participants to do their part.

Learning Team Charter

Course: ____________________ Semester ____________

Team Identifier: __________

All learning team members participated in the creation of this team charter and agree with its contents and its use to manage team interactions.
☐ (Please check)

Team Members:

Team Members	Phone	Email	Fax

Learning Team Objectives:

-
-
-
-
-

Potential barriers to achieving our team objectives we must be aware of and overcome:

-
-
-
-
-
-

Team Member Skills Inventory

Team Member	Skills	Team Role

Learning Team Charter

Ground Rules:

(Meeting schedules, locations, attendance expectations, assignment expectations, communication methods, meeting agendas, etc.)

____
____
____
____
____
____

Conflict Management:

What are the areas of potential conflict in the team?
How will we deal with the conflict? What steps will we take to manage internal team conflict?
What is the escalation process we will follow?

____
____
____
____
____
____

Faculty Review and Comments:
____
____
____

☐ Approved
Date: ______

Positive teaming is a program expectation and a career necessity. All team members are expected to fully participate in team's activities. There is no "I" in team, we want to encourage team-first attitudes so that the whole outperforms the sum of the parts.

FIGURE 4.2. Sample team charter, Seton Hall University (2008).

Learning Team Log

Team Identifier: ________ Date: _______
Course: ______________________________
Semester: ___________

Team Communications Activities
Meetings:
Conference Calls:
Chats:
Other:

Team Members:

Team Members	Participated
	☐Yes ☐No
	☐Yes ☐No
	☐Yes ☐No
	☐Yes ☐No
	☐Yes ☐No

Team Assessment for the Period
☐Exceeded our expectations
☐Met our expectations
☐Did not meet our expectations*

*Issues (what got in the way): ______________________

Team Actions (what did you do about it): ______________________

Lessons Learned (what have you learned): ______________________

Assigned Tasks:

Student	Task	Status

FIGURE 4.3. Sample team log, Seton Hall University (2008).

FORMING TEAMS

Instructors form teams in several ways. Sometimes, they allow students to decide among themselves. If teams were formed in previous classes, they may be populated again with the same participants, or the instructor may assign members to teams.

While I have tried many different ways to form teams, it turns out that when I assign members, there appears to be fewer conflicts. Following a simple formula, I wait about two weeks after the start of each course before assigning teams, giving me a chance to observe the talent pool and levels of activity. It's best to assign students from the same time zone in a team, if possible, but it's wise to select those who are not more than an hour's time zone difference from each other. This selection greatly reduces complaints from students about access and scheduling. While I have tried several ways of juggling talent—attitude, writing skills, work ethic, and content quality—unexpectedly, conflict has increased. I now ignore talent as an element in choosing teams and randomly assign students. In the end, I spread out skills, ensuring each team has strong and weak students. There is no full-proof method to team formation. Balancing strong students with weaker participants seems to help mitigate conflict to some degree and raises fewer evident issues because teams appear to be more balanced (Figure 4.4).

MANAGING TEAMING

Once teams are set, the instructor's role shifts to monitoring and encouraging participants. Because many team conflicts are

Setting the stage elements checklist	☑
Instructor teaming expectations clearly stated - Read Me First - "There is no "I" in Team	❑
Syllabus: - Clearly states teaming requirements - Assigns significant portion of the final grade to team assignments (15-30%) - Clearly states the teaming rubrics	❑
Plagiarism honor statement	❑
Learning contract	❑
Team charter	❑
Team logs	❑
Team formulation	❑

FIGURE 4.4. Setting the stage checklist.

often presented to instructors as "he said/he said" situations, it is challenging to manage intervention wisely. It's best to arrange for teams to perform in the open by creating team rooms to which you have access, allowing you to observe the team in action. Most course management systems offer this feature. Otherwise, gain access to team activities as a member.

It's also recommended that you restrict the use of e-mail in team communication. While participants commonly continue to conduct some of their business using e-mail, they should copy you in on their communications. Obviously, you will not be able to police all their e-mails, but students should be aware that that unless you have evidence of conflict, you will punish the entire team for dysfunction and grade accordingly. The threat seems to keep most activities in the open.

To support evidence of their participation beyond the team log, I recommend that students keep a personal record of their activities and encourage them to be as visible as possible in the team room. To show my hovering presence, I monitor team rooms from time to time—about once or twice a week—often posting a message asking if they need anything or are having problems.

INTERVENTIONS

Despite an active preventative policy, conflicts do happen from time to time, and interventions—soft, hard, and shock and awe—may be required (Figure 4.5).

Soft interventions are gentle reminders that the team needs to solve its own problems. When I see an issue developing or receive a call or an e-mail message from a team member worried about the team, I post or send a reminder about positive teaming behavior, the need to adopt a team-first attitude, and grade interdependence. On occasion, I will reach out to a particular student by phone to help mitigate conflict before it escalates. Soft

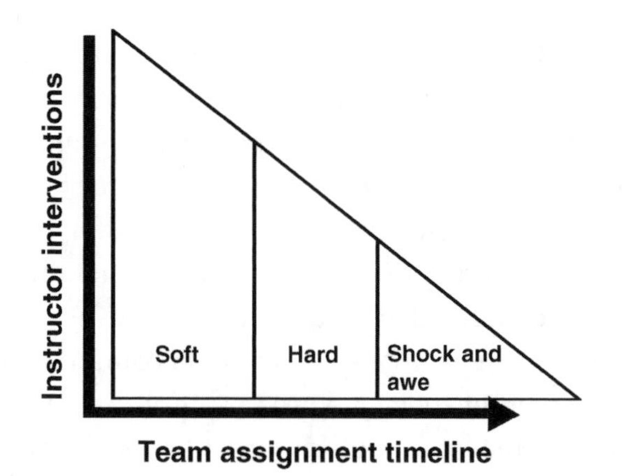

FIGURE 4.5. Instructor interventions matrix.

interventions tend to be delivered as advice or food for thought. You're not solving the problem but gently nudging students in the right direction, reminding teams that they are expected to recognize and resolve conflicts on their own.

CASE STUDY 1

An eager student jumped into the team's group pages before students were assigned to their teams and became frustrated by lack of interaction. She was further frustrated by the failure of newly assigned teammates to respond *immediately* to her demands:

> Greetings, Dr. Dool. I am not feeling very well because I feel like I am being ganged up on in regards to the team assignments. For starters, I have been working on the team assignments since before the class began. I devoted my entire weekend to getting all the information that is needed. I put my personal email address and personal phone number on the student lounge for everyone in the class to communicate with me before the class got off the ground, and since then I too received no communication. I just did not appreciate my team member trying to make me look like a irresponsible person who does not do her part, or the fact that she emailed me on (the school) email only.
>
> Student X

My soft intervention acknowledged her frustration and eagerness while suggesting she try to contact her teammate again.

> Student X, I am sorry you are having some early challenges, I appreciate you jumping right on the team assignment. Since it is so early in the class, this may only be a miscommunication. I suggest you contact them again and see if you can all work this out. Team assignments often have challenges and each member must find a way to turn it into a positive team experience. Let me know if there are any further issues.
>
> Dr. Dool

> Dr. Dool, I finally talked to my teammates and we have worked out the issues, thanks.
>
> Student X

Student X is an example of "Breathless in . . . ." She e-mailed me a few other times with other early warnings. Despite her anxiety, the team performed well and evaluations were positive.

If a soft intervention fails, you might take a direct approach with a hard intervention, such as telephoning a particular student, addressing the team as a whole, or changing team parameters. With hard approaches, it's wise to shift to specific recommendations to help teams overcome conflict.

Case Study 2

A recent team was troubled over a student who didn't fully understand how to act cooperatively. Earlier, in its charter, the team had adopted a majority-rules policy to resolve conflicts, but had not fully enforced it.

> Dr. Dool: We have a problem with X. He constantly argues every point and when we make a decision, he continues to argue. We agreed that majority rules but he will not honor it. This is really causing a problem. What should we do?
>
> Student X

The e-mail message below is an example of a hard intervention sent to all. I also phoned the student who was challenging the team's decisions. He explained that he felt his opinions were being ignored and wanted more input. He also agreed that he had not been as responsive as he should have been. I reminded him of his obligations and reviewed the team charter, which he had signed, emphasizing the majority-rules policy he and the other members had adopted.

> Team A—I am sorry you have had some issues so far in your team. It seems from what you have told me that the issues center on timely communication within the team and equal consideration of team member's suggestions. It also clear to me that despite the issues you have raised, you have delivered quality work so far. This tells me you are a capable team and are willing to make the commitment to do what is right. As I suggested when I spoke with some of you, every member need to engage with the team in a timely manner. You need to allow time to properly consider everyone's suggestion and you should share leadership responsibilities by giving everyone on the team the opportunity to lead on one of the deliverables. Let me know if you need any further help.
>
> Dr. Dool

No further conflicts were expressed by the team and they delivered assignments on time. However, other team members rated the student who was challenging the process quite poorly. Clearly, his early actions left some hurt feelings, but the team managed to overcome them.

If hard interventions fail, with learning objectives in jeopardy, it may be time for shock and awe. Taking action, it may be best to conduct a team teleconference. If the team appears to run ineffectively, you must be more direct and prescribe how it must conduct its business. Remind them that the team will suffer with poor grades if it cannot find a way to get its act together.

Case Study 3

A recent example uncovered accusations of plagiarism. One member accused another of plagiarism in a team room before any drafts had been submitted. While I did not intervene at this point, I continued to monitor the team room, witnessing several heated exchanges between two students while others remained on the sidelines. The conflict subsided for a week, when this e-mail appeared:

> Dr. Dool—Our graded assignment is due tonight and I do not know what to do. Student X's submission has clearly been copied from another paper. I told him this is not acceptable in our team space and he said he would correct it. He did change some sections but other sections were copied from another paper I found on Google. I am supposed to be the "submitter" but I do not want to submit it with his section in it but I also do not want the team's submission to be late. What do I do?
>
> Student X

My immediate response was:

> Student X, thank you for bringing this to my attention. If this is not original work, it should not be submitted, the entire team will be penalized. Please send me the paper, I will review it unofficially and then we will discuss how to move forward.
>
> Dr. Dool

The paper was clearly copied, and not very artfully. It showed up on the first page of my first Google search. Without accusing the student, I e-mailed the team, warning them their first draft was unacceptable. I gave the team a three-day extension (with an appropriate grade deduction) to resubmit its paper. I telephoned the offending student, explaining the rules concerning original work, the need to maintain integrity in team and individual deliverables, and the risks he was taking personally as well as the damage he might cause his team. I also reminded him that he had signed a learning contract and an honor statement. Finally, I also cautioned him that I scan submissions routinely for plagiarism using the Turnitin search software so it was not worth the risk. He offered some feeble excuses and said he understood.

In the end, I also required each team member to send me a personal e-mail message attesting that the work they delivered was either original or was cited properly. I warned them that if I did not receive signed documents from everyone, the team would fail the assignment. Everyone responded as directed, and there were no further conflicts. I did scan deliverables against the Turnitin database and, happily, found no further instance of plagiarism.

I wish I could say that these interventions work all the time, but sadly, some team dysfunction was so bad that it could not be resolved, ultimately leading to very poor team experiences, with team deliverables below expectation. Fortunately, these cases have been rare.

AFTERMATH

The biggest challenge in team assignments is awarding grades to individual students. Many instructors employ a one-grade-fits-all policy, with no distinction between team members. Others introduce ways to adjust grades based on individual contributions (see Chapter 3). I employ a team evaluation accounting (see the form below), requiring students to assess their own performance as well teammate contributions. It allows you to respond to legitimate cases in which students may have performed above average or taken responsibility for the work of others. I rarely adjust grades for students who are part of a team. In my experience, fewer than 10% of cases deserved to be adjusted and were certainly not without hard evidence. That's why I stress operating in the open (Figure 4.6). But giving yourself the option of adjusting grades can be useful. Opening the possibility of grade management can reduce student anxiety about team assignments and grade interdependence.

TEAM EVALUATION FORM

Your Name:	**Group:**
Activity:	

TOTAL POINTS FOR THE TEAM = (100 points per person X # of team members)
(Assign points to each category below 0 = Lowest Score per Category)

Group Work Self-Evaluation for: (your name)

1. Committed to the common purpose and consistently worked toward completion
2. Dealt with conflict constructively and was active in the decision-making process
3. Was cooperative and associated as a team
4. Was on time/not absent for class/team meetings and held themselves mutually accountable
5. Took initiative or leadership

TOTAL POINTS FOR GROUP MEMBER # 1

Group Work Evaluation for: (name)

1. Committed to the common purpose and consistently worked toward completion
2. Dealt with conflict constructively and was active in the decision-making process
3. Was cooperative and associated as a team
4. Was on time/not absent for class/team meetings and held themselves mutually accountable
5. Took initiative or leadership

TOTAL POINTS FOR GROUP MEMBER # 2

Group Work Evaluation for: (name)

1. Committed to the common purpose and consistently worked toward completion
2. Dealt with conflict constructively and was active in the decision-making process
3. Was cooperative and associated as a team
4. Was on time/not absent for class/team meetings and held themselves mutually accountable
5. Took initiative or leadership

TOTAL POINTS FOR GROUP MEMBER # 3

WORTH THE TROUBLE?

Because of early painful lessons in dealing with conflict in virtual teams, I introduced the process outlined in this chapter. While it adds to your workload and can be tedious, it does work. After deploying the full array of conflict mitigation steps outlined, evident conflicts in student teams decreased from over 60% to less than 20% and also led to higher student satisfaction survey results (Table 4.3).

	☑
Require that the team perform "in the open" using team rooms in the class site whenever possible. - Require instructor ability to observe	❑
Interventions: - "Soft" – (suggestions, nudging) - "Hard" – (specific recommendations) - "Shock & Awe" (prescriptive direction, direct intervention)	❑
The Aftermath: - Team evaluations - Individual and team grading rubrics	❑

FIGURE 4.6. Managing the process checklist.

TABLE 4.3. REDUCTION IN CONFLICTS AFTER EMPLOYING CONFLICT MITIGATION

Tracked Virtual Team Assignments	Evident Conflicts	Student Reaction: Enjoyed It and Said Experience Was Positive
127 Before conflict mitigation steps	78 (61%)	47 (37%)
98 After conflict mitigation steps	19 (19.4%)	57 (58%)

TABLE 4.4. CONFLICT MITIGATION IN STUDENT TEAMS

Sources of Conflicts	Mitigation Steps	Actions
General	*Setting the Stage*	*Specific Actions*
Dependence on technology	Clear and overlapping communication about teaming "expectations" and requirements.	Clear, detailed syllabus
Time management		There is no "I" in Team document
Team cohesiveness		Grading rubric
Trust issues		Honor statements
Lack of face-to-face elements		Learning contract
Language differences		Team charters and logs
Cultural, temporal, and spatial dispersion		Team evaluations
Specific	*Active Instructor Presence*	Instructor is in the online classroom every day if possible but at least five days per week, not missing more than a day in a row.
Differences in expected outcomes and commitment level	Evident presence in the online classroom	Instructor must be flexible and shift roles as needed from facilitation to intervention.

(continued)

TABLE 4.4 (*Continued*)

Sources of Conflicts	Mitigation Steps	Actions
Differences in the "quality" of contributions		
Differences in values		
Personality conflicts		
	Varying instructor roles including "traffic cop" and "cheerleader"	Teams should be expected to operate in the "open" through the use of "team rooms."
	Operating in the "open"	
	Interventions	Soft: gentle reminders suggesting the team solve its issues internally (e-mail, team postings, or calls)
	Soft	Hard: direct interaction through e-mails, calls or directive team postings
	Hard	Shock and awe: direct, live prescriptive actions
	"Shock and awe"	

SUMMARY

The conflict mitigation process requires setting the stage with proper expectations, a persistent and consistent instructor presence, and a willingness to intervene. It certainly adds to the instructor's workload, but results show it creates a more positive learning experience for both the instructor and the students. It also serves a higher purpose in that it helps prepare students to be positive, contributing teammates in their professional lives. Table 4.4 summarizes some of the key conflict mitigation actions necessary to reduce conflict in student teams.

REFERENCES

Bosch-Sijtsema, P. (2007) The impact of individual expectations and expectation conflicts on virtual teams. *Group and Organization Management*, **32**(3), 358.

Hinds, P. and Bailey, D. (2003) Out of sight, out of sync: understanding conflict in distributed teams. *Organization Science*, **14**(6), 615.

Hron, A., Hesse, F., Cress, U., and Giovis, C. (2000) Implicit and explicit dialogue structuring in virtual learning groups. *British Journal of Educational Pyschology*, **7**, 53–64.

Kaiser, P., Tullar, W., and McKowen, D. (2000) Student team projects by Internet. *Business Communication Quarterly*, **63**(4), 75–82.

Mortensen, M. and Hinds, P. (2001) Conflict and shared identity in geographically distributed teams. *International Journal of Conflict Management*, **12**(3), 215.

Sarker, S., Lau, F., and Sahay, S. (2001) Using an adapted grounded theory approach for inductive theory building about virtual team development. *Database for Advances in Information Systems*, **32**(1), 38–56.

Shin, Y. (2005) Conflict resolution in virtual teams. *Organizational Dynamics*, **34**(4), 331–345.

Smith, R. (2008) Learning in virtual teams: a summary of current literature. Retrieved September 3, 2008, from: http://www.msu.edu/smithre9/project12.htm.

Suchan, J. and Hayzak, G. (2001) The Communication characteristics of virtual teams: a case study. *IEEE Transactions on Professional Communication*, **44**(3), 174–186.

Volchok, E. (2006) Building better virtual teams. *eLearn Magazine*. Association of Computing Machinery (ACM).

CHAPTER

5

VIRTUAL TEAMS IN VERY SMALL CLASSES

ELAINE LEHECKA PRATT

All courses are designed with an optimum class size in mind. When introducing courses at universities, faculty consider not only the instructor-to-student ratio but also guidelines to stimulate productive group dynamics. With team projects, small-scale group dynamics may come into play, and the number of students in a team can figure prominently in class success, whether delivered in a live, face-to-face setting or online. At the Stevens Institute of Technology's WebCampus online unit, the optimum class size is approximately 20. Historically, some WebCampus programs began modestly, but over time (Figure 5.1), they have been building enrollments significantly.

I have been teaching WebCampus online courses since 2004. In two of my three classes, I assign virtual teaming projects. For my course on "Quality in Pharmaceutical Manufacturing," there are two team projects, and on "Regulation and Compliance in the Pharmaceutical Industry," there are three projects.

Virtual Teamwork: Mastering the Art and Practice of Online Learning and Corporate Collaboration. Edited by Robert Ubell

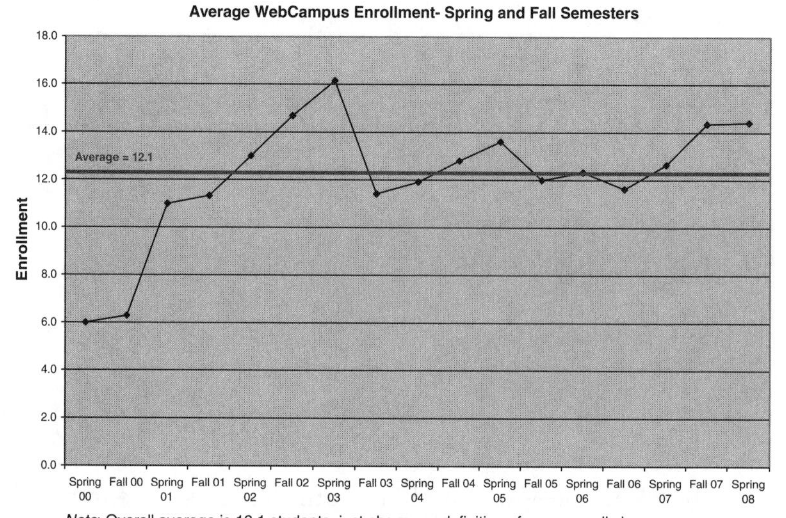

Note: Overall average is 12.1 students, just above our definition of a very small class.

FIGURE 5.1. Average enrollment each semester since WebCampus was launched in Spring 2000.

While initial enrollment in these classes was small, in recent semesters, there has been steady growth (Table 5.2).

Why would you offer a class with only four students? It may be part of a new program with low initial enrollment. By offering it frequently to accommodate student study plans, classes may be run with fewer students than is common. To permit students to complete their graduation requirements, it may be necessary to offer classes with modest enrollments. The class may be part of a corporate cohort, experiencing smaller-than-usual enrollment owing to work schedules or budget constraints. While there may be a number of reasons for teaching classes with few students, nevertheless, the important thing is to adjust them as needed to achieve success. Students in a small class should not feel cheated or overburdened. Herein lies the challenge of effectively managing virtual teams with small numbers of students in a class section.

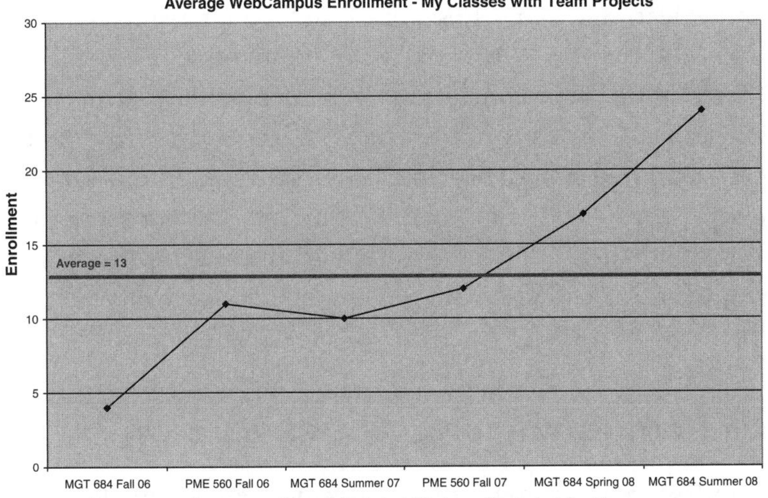

Note: Classes range from extremely small (4 students) to large (24 students), with an average of 13, close to our definition of very small classes.

FIGURE 5.2. Enrollment in the author's online classes with team projects.

GROUP DYNAMICS AND TEAM SIZE

The literature provides a great deal of published research on optimum team size (see also Chapter 3). Jeff Bezos, chairman, CEO, and founder of amazon.com, is quoted as saying, "If you can't feed a team with two pizzas, the size of the team is too large" (Fast Company, 2005). Research on team size goes back as far as the nineteenth century to French engineer Maximilian Ringelmann, who noted that the more people who pulled on a rope, the less effort each individual made (Mueller, 2006). The so-called Ringelmann effect is equally present in today's teams. In too large a team, some members can hide. In a recent *Fortune* article, "How to Build a Great Team," Jerry Useem (2006) says that 4.6 is the most effective team size. The University of Phoenix's *Learning Team Handbook* comments, "The size of groups is an important element in the success of online learning teams. Research with learning teams indicates that they work optimally with four or five members" (Betz, 2004). This conclusion correlates well with my own experience—teams of four or five seem to provide the best learning experience in virtual settings. But when you teach a class with four or five enrolled students, how can you build *teams*? Or if you are teaching a class of only six students, do you keep them together in one team only or break them up?

DESIGNING VIRTUAL TEAM PROJECTS

Before we review how to create effective teams in small classes, let's first look at virtual teaming assignments. Regardless of topic or course, virtual teaming gives participants practice in working together collaboratively on a short-term project designed to simulate an on-the-job experience (see Chapter 10). Working together in a team is part of the course experience and is as valuable as the topic. That is why it is important that the design of teams is carefully crafted and considered; a very small class must receive the same benefits from virtual teaming as a larger one.

In my classes, I require team assignments to be delivered as PowerPoint presentations on a given topic. In a traditional setting, the team presents slides in a classroom to other students. Online, the team posts its presentation to a discussion board where "live" commentary is added as notes. Since corporate work teams are often asked to present findings to an audience, the assignment gives students practice in critical presentation skills. A professional presentation is as important as covering the topic adequately. Consequently, I grade teams for both content and presentation. My virtual teaming assignments give students experience in three equally important factors: working in a team with those who they don't know who have varying backgrounds and levels of experience, acquiring in-depth knowledge of a given topic, and creating and presenting a professional summary of their work to a larger audience.

Before deciding how to establish teams in small classes, it is important to consider the purpose of your assignments. It is best that each team assignment possess a core theme (Table 5.1).

When I first designed the course, I projected an average class size of 16–20 students, with 4–5 teams per assignment. For some team projects, each topic is an essential part of the course. Assignments build on each other or show a progression. Eliminating some team topics because of small class size might leave

TABLE 5.1. EXAMPLES OF THREE TEAM PROJECT ASSIGNMENTS IN ONE OF MY CLASSES

Team Project	Assignment
1	*Regulatory Submissions—Requirements and Strategy*
	Team 1—Investigational New Drug Application
	Team 2—New Drug Application
	Team 3—Biologics License Application
	Team 4—Premarket Approval Application
	Team 5—Abbreviated New Drug Application
2	*Consequences of Noncompliance*
	Team 1—Barr Laboratories decision
	Team 2—Abbott consent decree
	Team 3—Schering-Plough consent decree
	Team 4—GE Medical Systems consent decree
3	*Consequences of Ethical Lapses*
	Team 1—Clinical Investigator Warning Letter
	Team 2—Division of Drug Marketing, Advertising and Communications Warning Letter
	Team 3—The New Lab Technician Case Study
	Team 4—Clinical Trials Outside US Case Study

Note: Each project has a central theme, although each presentation explores and develops one aspect of that theme.

out critical information. In other cases, where project topics are similar, eliminating one or two team topics does not matter as much. The point of the exercise can be demonstrated with only one or two teams (see the following cases for examples of each of these scenarios).

CASE STUDY 1. ESSENTIAL TEAM ASSIGNMENTS

In this example, the content of each assignment is vital. Students not only take away the team experience but are also required to view and study other team presentations for course content. In this case, the course is designed to cover all topics. If any of the teams have to be eliminated due to small class size, an integral part of the course is then missing. Table 5.2 shows different team assignments for a project in one of my courses. Students need to learn the material for each of these topics, so for a class with only two of the possible five teams, 60% of the subject matter is missing.

The five topics in Table 5.2 represent both progression and breadth, a scenario found in all of my courses, regardless of topic. Discussing multiple examples and viewpoints on a given

TABLE 5.2. FIVE TEAM TOPICS COVER THE PRINCIPAL PROJECT THEME OF REGULATORY SUBMISSIONS—REQUIREMENTS AND STRATEGY

Assignment
Regulatory Submissions—Requirements and Strategy
Team 1—Investigational New Drug Application
Team 2—New Drug Application
Team 3—Biologics License Application
Team 4—Premarket Approval Application
Team 5—Abbreviated New Drug Application

Note: Each topic covers a different Food and Drug Administration submission. With fewer than five teams, critical course information is not covered, so it must be covered in other ways.

theme is a classic pedagogical approach; in this case, it just happens to be paired with a virtual teaming exercise. In virtual teams, students benefit from being offered important information in different ways. To be fair and complete to small classes, all material presented in large classes must be covered. Ideally, team assignments cover all essential material, but with small classes, it is not always feasible or practical.

In a very small class with only two teams, it is best to select the two most critical topics. In teaching small classes, you will need to determine which topics are most common, most important, most challenging, or most interesting. Then, if you learn that enrollment in your class is low, you will have a plan in place to give your students hands-on experience of working in a team, researching the process, developing their management strategies, and crafting a compelling and persuasive summary presentation. After assigning the most valuable topics to your teams, you might post team projects that cover other topics delivered in previous classes so that your small class will still be exposed to all the material given to larger classes.

CASE STUDY 2. SIMILAR TEAM ASSIGNMENTS

An alternative to Case Study 1 is to assign multiple team topics on a theme providing depth of coverage. In this approach, eliminating one or more team topics does not irreparably gut the content of the overall assignment. This is an easier scenario for a very small class. For example, in one of my courses, a team project is to conduct an in-depth examination of a case study about a Malcolm Baldrige National Quality Award winner. The documentation is quite lengthy, so it is entirely possible for each team to examine just one section in detail. Depending on the size of the class, there are six possible subcategories of the application (Table 5.3).

TABLE 5.3. MALCOLM BALDRIGE NATIONAL QUALITY AWARD SUMMARY WITH SIX TOPICS FOR SIX TEAM PRESENTATIONS

Team	Topic
1	Section 5.1—Leadership
2	Section 5.2—Strategic Planning
3	Section 5.4—Measurement, Analysis, and Knowledge Management
4	Section 5.5—Human Resource Focus
5	Section 5.6—Process Management
6	Section 5.7—Business Results

In the example shown in Table 5.3, I created six topics for a large face-to-face class. In a typical online class, I might expect four teams. In a very small online class, there may be only two teams. For classes with fewer than six teams, you might prioritize topics in order of importance, matching topics to the optimum number of teams for that class. In this project, all topics are drawn from a single document so that even a small number of team presentations give the flavor of how the sample company prepared its application. In a larger class, students benefit from exploring more aspects of a company's quality system, but if a particular subtopic is not covered in a small class, the intent of the assignment is not lost. It might also be wise to post previous class presentations so that team members in a small class will have access to earlier deliverables. If you feel it necessary, you might also add course material on missing topics.

As the second case study illustrates, if you design multiple topics or outcomes into team assignments, it is easy to shrink the number of teams to fit the size of the class. However, a more serious challenge is presented when each topic is important to the learning experience and cannot be easily eliminated.

These case studies offer solutions for two teams. But what if there are not enough students for even two teams? What

TABLE 5.4. SUMMARY OF PROS AND CONS FOR DIFFERENT TEAM SIZES IN VIRTUAL TEAM PROJECTS

Team Size	Pros	Cons
1	Covers more topics in class	By definition, one is not a "team."
2	Covers more topics in class Small-scale team experience	Differences in experience, personality, or ability (or too much similarity) may hinder completion.
3	Simulates corporate teams Can subdivide assignments	May present "odd man out," if two side against one.
4	Simulates corporate teams Can subdivide assignment. Optimum team size for online class	N/A
5	Simulates corporate teams Can subdivide assignments	Some may hide or feel undervalued.
6 or more	Simulates some corporate teams	May be too large for online class. Some may hide or feel undervalued.

possible solutions are there? Your options are limited to assigning topics to individual students, creating smaller teams of twos or threes only, or introducing a single team of four with one topic and eliminating all others. There are pros and cons to each of these approaches (Table 5.4).

CHALLENGES OF MANAGING TEAMS IN SMALL CLASSES

An unsatisfactory solution for small classes is to assign individual students to each topic, solving the dilemma of covering all topics but, by definition, completely eliminating the teaming aspect. This solution also burdens students with work designed for four or more, which is not the point of the assignment. Since I also assign individual term papers, adding more individual projects is not the best approach.

Smaller teams can be an option. Assigning two students to a team may give them a sense of teamwork. Ideally, it can promote equal sharing, with leadership, division of labor, collaboration, and reaching consensus, among other qualities, occurring in miniature. While it may be a viable option, it often does not work out well, especially if the pair is of unequal experience, ability, or assertiveness. Frequently, one student tends to direct the activity, while the other does all the grunt work, coming away feeling undervalued. Pairs are also not representative of workplace environments. A team of three is yet another option, but it can lead to "three's a crowd," with two members siding against the other. With three, consensus is often difficult to achieve.

My favorite is a four-member team. It seems to provide enough work for all and allow for various viewpoints to be heard so that consensus can be reached without alienating anyone, yet it does not contain too many participants where opportunities are created for some to get lost or hide. There is a fine line between everyone in a team having enough work so that they are truly contributing versus the alternative in which participants feel that they are doing unproductive work or stepping on each other's territory, duplicating each other's efforts. One of my requirements is that each team include a summary of each member's contributions—who created which slides, who performed background research, who led the team, who compiled the final presentation, and so forth.

A team of five can also work effectively, especially for meaty cases with many requirements. Beyond five, however, you may encounter what Ringelmann called "social loafing"

(Mueller, 2006). Jennifer S. Mueller (2006), Wharton professor of management, who has done research on optimal team size, concludes, "Above and beyond five, you begin to see diminishing motivation. After the fifth person, you look for cliques. And the number of people who speak at any one time? That's harder to manage in a group of five or more." Since online teams communicate virtually—via chat or e-mail (see Chapter 7)—too many "speaking" at the same time is difficult to manage.

CREATING TEAMS IN SMALL CLASSES

So how can you build effective teams in small classes? Table 5.5 shows possible configurations for classes from 2 to 10 students.

TABLE 5.5. POSSIBLE TEAM CONFIGURATIONS FOR CLASSES RANGING FROM 2 TO 10 STUDENTS

Class Size	Team Configurations
2	1 team of 2
3	1 team of 3
4	2 teams of 2
	1 team of 4
5	1 team of 2 and 1 team of 3
	1 team of 5
6	3 teams of 2
	2 teams of 3
7	2 teams of 2 and 1 team of 3
	1 team of 3 and 1 team of 4
8	4 teams of 2
	2 teams of 4
9	3 teams of 3
	3 teams of 2 and 1 team of 3
10	5 teams of 2
	2 teams of 3 and 1 team of 4
	2 teams of 5

Note: Your assignment will dictate whether it is better to choose smaller teams or fewer larger teams.

If you opt for a small number of teams, it may be useful to post presentations delivered by previous classes, allowing students to view material covering topics that may be omitted from their current assignments owing to the limited number of options. Since students often like to compare their work in teams with present classmates ("Is our presentation as good/thorough/snazzy as theirs?"), introducing slides from earlier classes serves as a substitute in motivating students. Routinely, there is a significant rise in the professional style of presentations once students have viewed and received feedback about the initial delivery.

Let's turn to the composition of the teams and how they are formed (see Chapter 3). There are two principal ways to form virtual teams—either by student selection or by instructor assignment. Typically, my online classes are composed of a number of part-time students who are also working professionals, some full-time graduate students, with varying degrees of work experience, and a few full-time undergraduate students, with little or no or workplace history. Some are drawn from the same company, and a few full-time graduate and undergraduate students often know one another because they have had classes together before. Recognizing that previous personal connections may give some teams advantages over others, I avoid allowing students to form their own teams. Instead, I assign students to teams. I also change team composition with each new assignment.

When initial team assignments are made, students and faculty rarely know each other very well. So it is impossible for instructors to judge who might be a good team member and who might be a slacker. For your first team, it is best that you assign members randomly; alphabetically, for example, with a mix of working professionals and full-time students. In subsequent assignments, you might vary group composition so that students experience team members with different personalities, skills, and background. In forming teams later, once you have a better sense of who the excellent students are and who are less resourceful, you might take these qualities into account.

Other instructors establish teams at the beginning of each semester, maintaining the same composition throughout the course. This approach may be valuable, especially when team projects build on each other, yielding a substantial final product. On the other hand, if your assignments are designed to create short-term experiences, it might be best to form teams composed of members with varying skills and background who may not know each other and who come together quickly to complete a particular task. If students can master this skill, they can probably work together skillfully on continuing projects with long-term goals. It may also be useful to give students opportunities to play new roles with each new team. For example, if a student is not a leader in one group, he or she may emerge as one in the next. Keep in mind that while I prefer to watch leaders assume their roles naturally, some instructors assign leaders to teams. Finally, if team members initially get stuck with a slacker, it is not fair for them to have to continue to carry him or her all the way through until the end of the term.

TEAM DYNAMICS AND INTERACTION

In the Stevens Institute of Technology's WebCampus, there are a variety of ways for teams to meet virtually (see Chapter 7). Some communication software is built into the student learning management system, such as e-mail, discussion board, and chat rooms (Figure 5.3). Teams are also free to set up their own teleconferences or even arrange face-to-face meetings.

When I assign a team project, I open chat rooms for each team (Figure 5.4). The chat room is frequently used as an initial

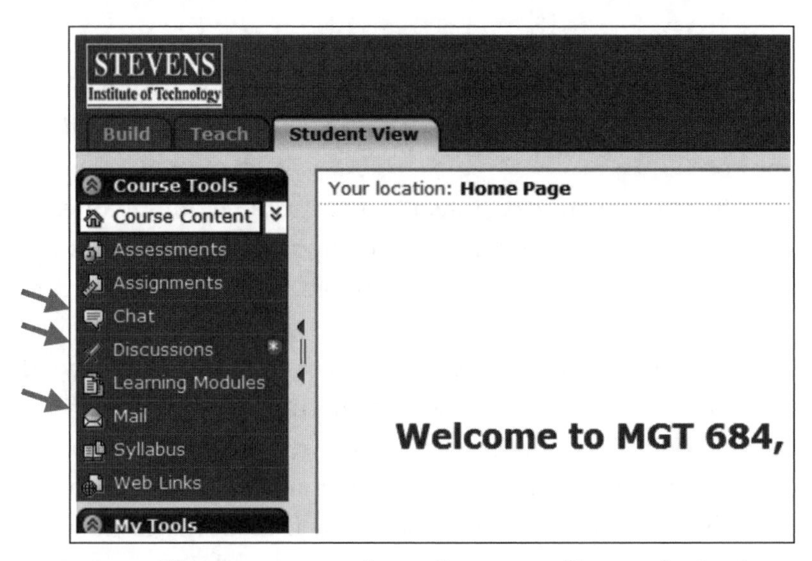

FIGURE 5.3. Homepage menu of an online course. Communication functions, such as chat, discussions, and mail, are highlighted.

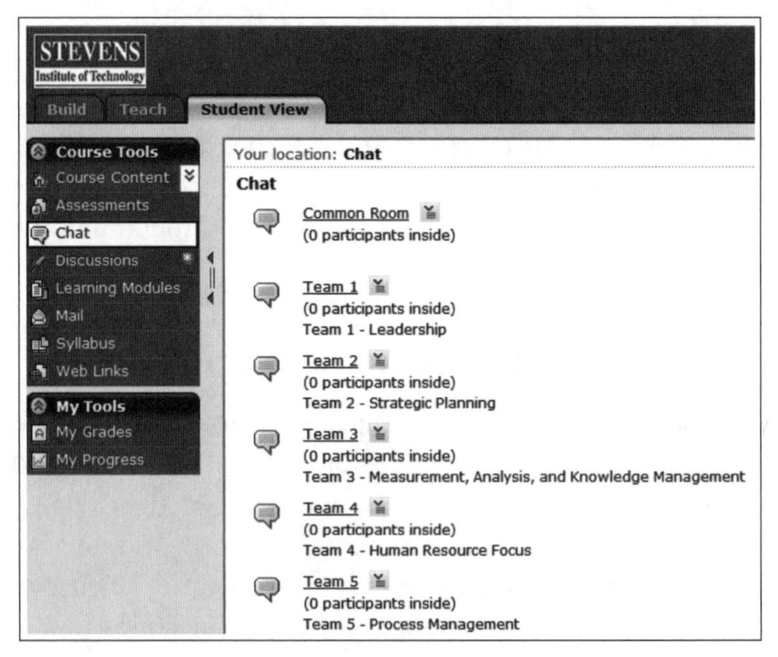

FIGURE 5.4. Chat rooms for the Malcolm Baldrige National Quality Award (discussed in Case Study 2). Teams use chat to discuss projects in real time.

way of getting together as a team. Afterward, teams may set up teleconferences and other ways to meet.

Personality and style play significant roles in team assignments. Typically, one or more team members will take control as the leaders, establishing schedules, handing out assignments, managing contributions, and arranging draft presentations (see Chapter 2). Occasionally, leaders may become overbearing, forcing others to rebel. When more than one strong member seeks to take control, there is often a clash. In some classes, a few team members—for whom your class is their whole life—assume control without being asked by others in the team who are not as free and who are commonly fully employed in demanding jobs in industry. It is also challenging when the team is composed of some from abroad whose first language is not English. Others may resent having to deal with English grammar skills (see Chapters 8 and 9). There are also loafers who expect the rest of the team to carry their load. Certainly, these obstacles are present in classes of any size, but they are often magnified in very small classes because there is nowhere to hide.

If teams come upon concerns that they cannot resolve on their own, I encourage them to e-mail me privately (see Chapter 4). The two most common complaints are about overbearing leaders or slackers not pulling their weight. In addition to requiring information on each member's contributions, I also ask students to grade each member of their own team. These confidential evaluations are very valuable to sort out differences between members in a team. In my classes, the final grade is usually the average of the individual and team assessments (see Chapter 3). When students realize that their individual contributions as well as the final group assessment will be taken into consideration in their grade, they often become thoroughly involved and vigilantly guard their contributions and make sure they are reported accurately.

Whenever there is conflict about accurately reporting contributions made to the team effort, it is best to encourage the team to work it out among themselves. If accusations continue, it is a good idea to require corroborating evidence from the team before reaching a conclusion. These situations often mimic real life. Employees on corporate teams must also deal with strong personalities and slackers. Fortunately, these problems are relatively rare.

BENEFITS OF VIRTUAL TEAMING IN VERY SMALL CLASSES

What are the benefits of virtual teaming in very small classes? If you craft a well-thought-out small class "Plan B," the student experience should be quite valuable. With fewer students, you may have more time to give detailed feedback on each assignment. But most importantly, from the student's perspective, a small class (and perhaps a smaller team) gives the student a chance to shine. It is well known that if you teach something, you generally learn more than your students. So in addition to the experience gained in working in a team and delivering a professional presentation, students also learn assigned material by researching it well and communicating it in a clear and professional manner. With each new semester, I find that students almost always seem to uncover new facts, ideas, or ways to present information.

SEVEN TIPS

Here are some distilled suggestions you may find useful:

- *Tip 1. Do not sacrifice team goals to cover topics.* Make sure that you form teams that will provide your students with a robust virtual teaming experience, even if this is best achieved with only one or two teams in a very small class. As important as the content is, it is equally important for the process itself to turn into worthwhile learning experience.

- *Tip 2. Prioritize team topics.* By prioritizing, you learn which topics must be assigned to deliver an uninterrupted flow of course material and which might be considered reinforcement and thereby optional for small classes. If there are not enough students in your class to cover essential topics adequately, post former student presentations to illustrate essential material or cover it yourself during lectures.
- *Tip 3. Keep the number of members in each team as equal as possible.* If the numbers are divided more or less equally, each student has a similar workload. It is unfair, in a class of 10, for example, to create one team of 4 and 3 teams of only 2 each.
- *Tip 4. Provide small teams with extra help.* If a very small team is struggling, step in and give suggestions about how to manage the workload. You do not want the team to be so overwhelmed that it will fail to grasp the content of your assignment.
- *Tip 5. In courses with multiple, unrelated team projects, assign different teams for each project.* This approach simulates teaming in corporate settings in which members serve on several concurrent or successive teams. It also mixes up leaders and loafers, and those with varying degrees of experience. However, in those courses in which projects build on one another, it is best to maintain the same team composition throughout the course.
- *Tip 6. Make sure your instructions and grading criteria are clear.* Make your expectations clear at the start of each team project, outlining the scope of each assignment and

what questions need answering. Provide parameters for team presentation, such as using large, clear fonts and additional commentary as notes. Require that students identify who prepared the material for each slide. When grading, make sure you alert students to cover their topics completely and deliver their presentations professionally. Also make sure you assign a due date for team deliverables. In my classes, I grade individually and by team, with the final grade being an average of both.

- *Tip 7. Offer quality feedback after each team project.* Virtual team projects are most valuable when you give quality feedback, commenting on content, presentation, notes, sources, and division of labor. Students appreciate praise and absorb constructive criticism.

In online courses, virtual teaming is a valuable exercise. Simulating workforce teams, students practice critical skills in a controlled atmosphere. Students are exposed to a wide diversity of backgrounds, experiences, enterprises, and cultures. I am often amazed at the level of creativity and professionalism that my students bring to their assignments. Many enjoy learning from each other's experiences, and they appreciate the opportunity to work with different groups. In very small online classes, your first inclination may be to discard virtual team projects, but as this chapter has shown, with careful planning and management, virtual teaming can be successful in small classes as in classes of any size.

REFERENCES

Betz, M. K. (2004) Online learning teams: indispensable interaction. *International Journal of Instructional Technology and Distance Learning*, **1**(6), 3.

Fast Company, editor (2005) *The Rules of Business: 55 Essential Ideas to Help Smart People (and Organizations) Perform at Their Best*. Doubleday Business, p. 159.

Mueller, J. S. (2006) Is your team too big? Too small? What's the right number? Knowledge@Wharton, June 14.

Useem, J. (2006) How to build a great team. *Fortune*, June 1.

PART

2

Virtual Team Technology

CHAPTER

6

CHOOSING ONLINE COLLABORATIVE TOOLS

PHYLISE BANNER, M. KATHERINE (KIT) BROWN-HOEKSTRA, BRENDA HUETTNER, AND CHAR JAMES-TANNY

Selecting the right tools for your online team starts with determining which ones are likely to work best for what you have in mind. Even though people may say, "We need X," they can easily choose solutions that more closely reflect current fashion than what is really needed. To pick the right solution, you must first identify the problem you are trying to solve, along with several other relevant pieces of information, before deciding which application is best.

Your first step in achieving your objective is to perform a "needs analysis" to appreciate the problem you are seeking to solve and then identify the tasks that must be accomplished to reach your goal. Just as you wouldn't use a hammer to cut a board in half, so too you wouldn't select a chat program to create course content. Most projects require a variety of tools for different purposes. A good way to begin is to determine what needs to be created, which team members will

Virtual Teamwork: Mastering the Art and Practice of Online Learning and Corporate Collaboration. Edited by Robert Ubell

be responsible for working on it, and who will be receiving the end product.

NEEDS ANALYSIS

A needs analysis covers these key steps:

1. Determine what problem you are trying to solve.
2. Identify your current status, using a co-called "SWOT" (strengths, weaknesses, opportunities, threats) analysis.
3. Inventory current capabilities (tools, processes, team, and so on).
4. Identify your goals.
5. Recognize the gap between where you are and where you want to be.
6. Identify communication needs and types.
7. Compile your selection criteria and identify tasks.
8. Determine your budget and schedule.

Even if you do a good job performing your needs analysis, there are a number of hazards facing you as you select what's best. Before you start, it's good to be aware of a few obstacles that may be in your way. Vendors constantly update their applications, and new ones are being continuously created. Since new technologies enter the marketplace at a fairly rapid rate, the selection you make may become obsolete or dated even while your project is still underway. Accessibility varies, so while you may imagine that what you choose is universally accessible, it may not operate on everyone's computer or in every environment. Also be aware that many collaborative applications take place in the "cloud"; that is, they are web-based.

What Problem Are You Trying to Solve?

Start by identifying the problem you want to solve. The table below outlines some common objectives and who might be given the assignment to contribute to the solution.

Output	Who Contributes?
Schedules, budgets, plans	Instructor, department head, or designated team member; possibly all members
Templates, standards, guidelines	Instructor or external source
Team discussions	All members
One-on-one discussions	Single member
Status reports	Individual members or one member for the group or subgroup
Content	All members

What is Your Current Status?

A SWOT analysis helps you identify technical and process issues that may affect tools you select.

- *Strengths* are inherent advantages in solving problems; for example, your students may be technically savvy.
- *Weaknesses* make it harder to solve problems; for example, some students may have dial-up connections.
- *Opportunities*, outside your control, may make it easier to solve your problem; for example, your helpdesk is exceptionally cooperative.
- *Threats*, also outside of your control, make it harder to solve your problem; for example, your IT department requires you to employ an application that no one in your class knows how to use.

While a SWOT analysis can often be performed fairly quickly—you may already know what you will be facing—it's best to jot down your results anyway to maintain a baseline. When analyzing multiple variables and multiple scenarios, it's easy to become confused by the details, if you don't document them. As you identify each category, it's good to consider the relative importance of each item, as well as its impact, if something goes wrong. This exercise will help with risk analysis later on.

WHAT ARE YOUR CURRENT CAPABILITIES?

In industry, project managers build teams by choosing people based on their experience or knowledge. In an online class, teams may be formed by self-selection, randomly, or by assignment. Instructors rarely know the level of knowledge in their online class until the course begins—and then it's often too late to change to a more widely known tool if students are unfamiliar with the one you select. When introducing a new set of tools, it's prudent to consider your own level of knowledge and comfort, the likelihood that your students will know how to use them, ease of use, and the availability of training and technical support.

At the start, inventory hardware and software your campus supports. Some collaborative tools may already be available at your institution. With just a basic Internet connection, for example, you can access tools such as voice-over-Internet phone tools, which include Skype, and most instant-messaging packages—Yahoo, AIM, or Meebo. If your college or university uses a learning management system (LMS), such as Blackboard, Moodle, or Desire2Learn, it's wise to explore its capabilities. It's also useful to review these technical details:

- *Bandwidth.* What speed do faculty and students use to connect to the Internet?

- *Disk space.* Does everyone have enough disk space? Is there enough space on the server, if you are using an institutional system?
- *Equipment.* How old is everyone's equipment? What operating systems are in use? Are computer labs available for those who don't have their own?
- *Software.* What software does everyone have installed? Which versions? Is there any incompatibility?
- *Browser requirements.* Not all browsers are supported by all tools. Be sure to specify which browsers should be used by students when accessing your online classroom.
- *Security protocols.* What security systems do faculty and students have in place?
- *Internet, intranet, virtual private network (VPN) access.* Can everyone access the file server and network off-site?
- *Conference calls.* What systems are in place? Does your institution provide access? If not, can everyone agree on the same application? (Chat and other/instant-messaging packages can be used for conference calls, although there are limits and bandwidth constraints in some areas.) For international students, is there a toll-free, long-distance number available?
- *Support services.* What services are in place for students at a distance? What hours is your helpdesk open?
- *Assessment.* Is there an assessment rubric in place? Are students aware of how their participation will be evaluated?
- *Knowledge management and institutional governance.* Who retains rights to collaborative work?
- *Copyright compliance and plagiarism.* How do you plan to avoid honor code violations?

Learning how to use collaborative tools is rarely the main focus of an online course. Still, it's prudent that you not only provide instruction about concepts but also give your students the ability to navigate applications successfully. Particularly when introducing new technology, it's important to build flexibility into your curriculum to give students a chance to become familiar with it.

In addition to accommodating tools and technology, you may need to modify assignments in other ways. While virtual tools allow you to communicate with participants from most parts of the world, you still need to keep track of where everyone is. A synchronous learning activity scheduled for 3:00 p.m. in Boston may not be too difficult for students to attend who reside in London (where it would be 8:00 p.m. that evening), but it might present a serious obstacle for students in Hong Kong (where it is 3:00 a.m. the next morning).

Online learning can present other technical obstacles because students and faculty are often scattered all over the world with different levels of technical expertise. Many in Africa, for example, and others in parts of Asia and South America, often lack infrastructure to support continuous broadband connectivity. Many parts of the world frequently experience rolling blackouts and other service disruptions that almost never affect students in North America and Europe. War and other conflicts can also play havoc with technology and infrastructure.

WHAT ARE YOUR GOALS?

Once you have identified your current status, it's time to think about what you want to accomplish in your online class and which tools to employ. How does online learning support your learning objectives? Why does your course need to be taught online and how will the virtual environment affect students' ability to assimilate knowledge?

As you consider these questions, it's useful to think about how to accomplish your educational objectives. How does the virtual classroom allow you to execute your course goals, learning objectives, and assignments? How will your learning outcomes be achieved? Remember to employ SMART (specific, measurable, achievable, relevant, time-bound) goal setting.

THE GAP BETWEEN WHERE YOU ARE NOW AND WHERE YOU WANT TO BE

Once you perform a SWOT analysis, you will see more clearly where you are now, compared to where you want to be. In a traditional classroom, you might have introduced assignments that achieved your ends successfully, but they may not be as effective online. Or, you may discover that new technologies present opportunities to explore course concepts in new and better ways.

Mapping where you are against your goals helps to plan for how to get from where you are to where you want to be. It also helps to prioritize tasks and rank them in importance. Introducing a new technology can be frustrating and time consuming. Effective planning and analysis at the start helps prevent problems that confront every project. If migrating to a new technology is likely to be complex, consider a phased introduction in which you pilot the new tool first in one class, work out the bugs, and refine it before offering in all your courses. It gives you a chance to take a step back to determine whether it's worth the effort.

COMMUNICATION NEEDS AND TYPES

Most classrooms have similar communication needs. The following is a list of the most common ones. It may be useful to identify additional communication options relevant to your course.

- *One-to-many.* Typically, instructors deliver lectures or provide information about course concepts or give instruction covering assignments. Online, instruction can occur in a number ways—in webinars, printed materials, web sites, podcasts, message boards, chat, and so on.
- *One-to-one.* Instructors frequently hold conversations with individual students. Online, such interaction can occur employing e-mail, or by phone, instant messaging, and so on.
- *Many-to-many.* Classroom discussions, team projects, and group presentations are examples of many-to-many communication. They can be held using e-mail, message boards, chat, webinars, collaborative tools, and so on.
- *Many-to-one.* Team assignments often require members to meet with the instructor to respond to questions. Chat, message boards, and e-mail can be effective tools for these communications.

SELECTION CRITERIA AND TASKS

In the software world, selection criteria are also known as requirements, features, or functionalities that support the results you want to achieve. For example, if you are looking at a complete LMS, you might want to integrate a grade book, a calendar, and testing tools, among other applications. If your institution already provides an LMS, be sure to learn whether the collaborative tool you are considering is interoperable.

Consider the tasks you need to perform in your online class and jot them down. It helps to prioritize how important particular features or functions are in order to achieve your objectives. Then, compare available tools with what you plan to do. Consider these common instructor tasks:

- Establishing rapport, setting expectations, setting a tone
- Providing course content
- Assigning homework
- Returning graded assignments
- Posting grades
- Testing student knowledge
- Responding to questions

Also consider these common student tasks:

- Turning in assignments
- Communicating with teammates on group projects
- Participating in class discussion
- Studying for examinations
- Communicating questions with the instructor

It's also prudent to consider other factors that can contribute to effective deployment, such as how usable the tools you select are. What is the cost to your department or to students in your online class for the application you have in mind? It's also wise to consider how long it may take to train students. Will the introduction of the tool and student training fit within the semester? Have you explored whether other classes have used the tool you are considering and, if so, whether they experienced difficulties? Will your IT unit support it with a helpdesk? Is there a tutorial available?

BUDGET AND SCHEDULE

Even though many tools offer educational discounts or may even be free in academic settings, some still may require time and effort to install. Deployment can be expensive if what you have selected will need significant time commitment to introduce. To get buy-in, it's helpful to know not only the cost of the application itself but also how much time it is likely to take to install. Knowing how long it will take to set up will help you devise an implementation schedule, with allocations for training students.

It's a good idea to build in extra time in your course schedule for everyone to get up to speed. The first time you hold a synchronous class meeting, some of your students may be unable to connect. If you encourage instant messaging, for example, take the time to ensure that everyone on your team recognizes each other's login names.

CREATING A BUSINESS CASE

When making a proposal to your managers or administrative deans, it is important to provide a cost-benefit analysis to justify your idea and the resulting expenditure. Use these guidelines:

- Describe your idea and why it will benefit the university. Be as specific as possible, including costs, hours, and personnel requirements as appropriate.
- State the pros and cons listed as a bulleted list. Relate your plan back to the university's strategic plan or your department's goals wherever possible.
- Explain how and when it should be implemented.
- Offer alternatives, if possible.
- Request action from the department or dean, if needed. Be sure to include due dates and action items, if appropriate.

- Answer questions that the department or dean might have about implementation—for example, How much staff/volunteer time may be required? Is there an initial cost to the university? How much?
- Provide your full contact information.

TOOL CATEGORIES

Categorizing tools into common groups makes it easier to find the right solution. Consider these:

- Collaborative software suites
- Chat and meeting tools
- Information broadcasting
- Information sharing
- Information gathering
- Project management
- Wikis
- RSS feeds and other "push" technologies
- Learning management systems
- Social networking
- Bibliographies

COLLABORATIVE SOFTWARE SUITES

While known also as "groupware" or "computer-supported cooperative work," there is a fine—but important—distinction between cooperation and collaboration. Cooperation is implied when the results of independent subtasks are merged to create a final product. Collaboration requires teamwork for each subtask as members work toward a common goal. Collaborative software suites offer many features also found in other categories, such as e-mail, forums, or chat. Applications such as Drupal, Plone, and TikiWiki fall into this category.

CHAT AND MEETING TOOLS

Applications in this category are used to communicate with team members.

CHAT

Chat tools can be external, internal, or system based. Some institutions do not permit external tools to be installed on campus systems because of vulnerabilities (such as viruses that may be transmitted through chat programs). Other institutions allow internal server-based tools because the environment can be controlled. Web-based tools use a browser (such as AIM Express or Meebo) or Java applet (such as Jabber).

Some common chat programs offer connections to applications such as MSN, AIM, Yahoo, Skype, Trillian, XChat, ICQ, Jabber, and GAIM. Most chat programs allow you to set up a chat room for multiple users. Trillian, a multi-protocol application, allows you to connect to AIM, ICQ, MSN Messenger, Yahoo Messenger, and others and permit chat rooms with the same protocol only (i.e., you cannot install a chat room with both AIM and MSN contacts). Meebo, a web-based multi-protocol application, permits mixed-mode chats. If your institution has adopted an enterprise-wide LMS, it may have built-in access to chat and meeting tools.

Skype uses a voice-over-Internet protocol (VoIP) that gives your computer a high-speed telephone-call connection. You can call other Skype users without charge to make and receive calls. Skype's conference-call option allows up to five participants. However, you may find that VoIP applications are unreliable if you run several other tools at the same time. Be sure to check memory, bandwidth, and system recommendations.

MEETINGS

Web-based applications, such as WebEx, GoToMeeting, NetMeeting, and a wide variety of other commercial and free software, give you the ability to hold meetings, conferences, and training sessions online. Applications are distinguished by their pricing structures and limitation on how many may meet at one time. Some meeting tools operate with a telephone bridge and may not connect internationally. An alternative is Skype, since calls to other Skype members are free. In place of these options, most LMS offer synchronous communication tools that can be used for online meetings.

INFORMATION BROADCASTING

Applications in this category are employed to distribute information to team members and others with blogs, webinars and presentations, and podcasts.

BLOGS

Blogs (short for "weblog") are online diaries that combine text, images, and hyperlinks. Blogs allow you to categorize entries and limit visitors to enter posts on specific topics only. Some blogging software permits multiple bloggers to post entries, either to the same blog or to multiple blogs linked together (known as a "blog farm"). Blogs have emerged as the most common web tool. In correspondence schools, "journaling," a blog cousin, emerged as an effective distance-learning activity.

Blogs can be hosted on a blogging software server or can be installed on an institution's server. If the URL refers to the blogging software's site, the blogger is using that server. If the URL refers to the blogger's home page or domain, the blogger has installed the software on a personal server. The difference between hosted services and self-hosted blogs is determined by who performs maintenance. Hosted services manage all maintenance, including updates. Certain LMSes may have blog applications built in.

WEBINARS AND PRESENTATIONS

Webinars and presentations give online students the ability to be "present" in a virtual class, exploiting tools that are largely the same as those noted in "meeting tools." Used in a manner similar to training sessions, the primary difference is interaction. Typically, in a webinar or presentation, there is limited interaction.

In industry, teams regularly conduct webinars with multiple presenters. Presenting online is significantly more challenging than delivering talks in person because of the loss of visual cues. To engage the audience, many webinar programs offer chat and "hand-raising" features. Some presenters leave a phone line open, but calls during presentations can be distracting.

Some instructors encourage students to take command of delivering webinars, recognizing that students with online presenting skills can be valuable in industry. Most systems allow webinars to be archived so that students can retrieve presentations at their leisure. It gives students, unable to attend the real-time webinar, the chance to view the presentation at a later date. Archiving is also useful to students who want to review material they may not have fully absorbed during real-time presentations.

PODCASTS

While presentations or webinars are typically scheduled, podcasts allow team members to retrieve information on demand. Podcasts deliver audio files, while vodcasts show videos. With very little equipment, users can easily create a podcast for distribution. To create a podcast, you will need a microphone and recording software. Audacity is the most frequently recommended software for recording podcasts, as it is open source, cross-platform, and free. To create a vodcast, you will also need a camera and video editing software. Applications such as Hipcast are often used to create audio blog entries (basically a podcast incorporated into a blog).

Podcasts are often available for download to iPods, MP3 players, and other mobile devices, a very accessible way of delivering lectures. Encouraging students to create podcasts is also a good way for them to share their work with the rest of the team.

INFORMATION SHARING

For virtual teams to be effective, members must share information. Many collaboration software suites include calendars, file galleries, forums, bulletin boards, application sharing, content management, and workflow management, all of which can be used to share information. However, you can use other programs for each purpose.

INFORMATION GATHERING

Surveys, project management and scheduling, feedback, and time tracking allow team members to collaborate on current assignments.

SURVEYS

Surveys allow you to collect information, usually without verification, about a specific activity or concept. They help participants understand an activity or concept, identify areas of concern, and determine whether more information needs to be gathered. To ensure scientific validity, it's wise to employ quality sampling techniques.

FEEDBACK

Feedback forms and polls can identify areas where course content may not be clear. It can also identify areas where you can improve course logistics, content, and assignments. Use feedback to adjust your project plan and identify topics for future surveys.

PROJECT MANAGEMENT AND SCHEDULING

Project management and scheduling allow you to monitor progress governing specific tasks. These applications also allow you to closely follow student assignments. Project management software can pinpoint both the strengths and the weaknesses in a team. They are effective in helping students manage assignments and semester-long group projects and are also useful in helping you plan and schedule your courses, especially when integrating new technology in your curriculum.

TIME TRACKING

Time tracking, often considered part of project management, permits you to judge how much time is being spent on a specific task. Most LMSes have built-in student time tracking options that allow you to see how much time a student is spending on the course and on your assignments.

WIKIS

Wikis are collaborative web sites that allow users to add and edit content. The word "wiki" can refer to the site or the software installed on the site. Wikis are similar to word processors, but use specific codes to add or modify content, facilitated by a browser in real time. Most wikis disable JavaScript and HTML tags, which help to keep the results fairly secure. Some wikis have been enhanced to include content management capabilities, such as user permissions, categories, and other actions.

While wikis can be installed locally, they are almost always run from a server (either Internet or intranet). Wiki farms provide hosting for individual wikis. Most wikis include these features:

- *Registration.* Some wikis require registration before content can be added or edited. Others are open (and require constant monitoring because of hackers).
- *History.* A tracking application allows you to view changes over the life of the page and gives you the option to revert to a previous version. You can usually choose versions you want to compare.
- *Locking pages.* A wiki administrator can lock specific pages against changes. It's best to lock pages that only an administrator can modify, such as your home page.
- *Search.* Some wikis let you search titles only, but many include full-text search.
- *Recent changes.* These pages are those most recently changed. A quick glance at any recently changed pages indicates which ones may have been spammed.
- *IP blocking.* An Internet Protocol (IP) address is the numeric equivalent of a URL, the code used to identify a

computer network. The IP blocking feature in many wiki packages prevents visitors from specific IPs from accessing the site. An administrator must add IPs to the list, although this is usually a moot exercise (hackers change their IPs too frequently for a lock to work well).

For a complete list of wikis, go to Wiki Matrix at http://www.wikimatrix.org. You can use the comparison chart on the site to determine which wiki is best for you. Most wiki applications provide special offers for academic institutions and integrate (using APIs and other system plugs) into your LMS.

RSS Feeds and Other "Push" Technologies

"Push" technologies deliver information from a server to users. Users may initially request the information, but it is delivered when ready. Users have no control over when the information arrives. "Pull" technologies, such as web sites, are retrieved when the user requests. Push technologies include e-mails, faxes, voice mails, newsletters, and really simple syndication (RSS) feeds.

If your team has set up items in different categories, team members can receive notifications by signing up for RSS feeds. RSS aggregators send notification when content on particular sites has been updated, so that you don't have to visit every site to see if anything has changed.

Learning Management Systems

Systems of this type are also known as course management systems (CMS), collaboration and learning environments (CLE), or virtual learning environments. No matter what it's called, very likely, your college or university already has one in place. As you consider various technologies to use in your online class, make sure that the tools you select can be seamlessly integrated with your LMS. Most LMS products include a variety of common features, such as document distribution, a grade book, live chat, assignment uploads, and online testing. It's best to consider a system that successfully integrates blogs, wikis, RSS, and other collaborative tools. The principal ones on the market are Angel, Blackboard, Desire2Learn, eCollege, Moodle, and Sakai.

SOCIAL NETWORKING

Social networking and social media web sites, such as Twitter, FriendFeed, Facebook, LinkedIn, and Delicious, provide alternative ways of keeping in touch with team members. Some sites allow you to create groups, and some have limited security settings. But before your team engages in social networking, it's wise to consider social-networking privacy policies, if you don't want others to see your work and conversations.

BIBLIOGRAPHIES

When composing research papers, you can save time and increase efficiency and accuracy by using a bibliographic application, such as Zotoro, EndNote, and WizFolio. Most of them allow you to copy and paste bibliographic information directly into the application and also allow you to take notes about the reference.

TASKS

After you have a clear picture of the learning objects you want to create, your current capabilities, and the needs of your faculty and students, you can begin to evaluate the collaborative tools you may want to introduce. Your choice will depend on many factors, but the following list of tasks (parts of which are excerpted from Brown et al., 2007) gives you an idea of some of the possible approaches.

USING TECHNOLOGY

When migrating to a new technology, remember to build training time into the first part of your course. You may incorporate training into an assignment that provides a core concept. If you are exploiting the tool in a novel way, or if it is an application that most students are unfamiliar with, be sure to note resources where technical support can be found. Consider these options:

- Web site with links to online syllabus, tutorials, FAQs, and other resources.
- Contact information for technical support
- Wikis and forums where students can share experiences

ASSIGNMENTS

For documents, such as a presentation, outline, research paper, or other class assignments, multiple parties are likely to contribute to the products as authors, editors, or approvers. The output is also likely to go to many recipients. To accomplish these tasks, here are some options:

- Wiki
- Google Docs
- Zoho
- Forums or message boards
- Web sites

- Blogs
- Shared team spaces

STATUS AND TEAM NOTES

Typically, status reports are created by a single team member and are distributed to others. They may also cover what other team members are doing. Generally, status reports are archived by date, with each subsequent one stored as an additional file, a system that records efforts that may have worked well and others that didn't. Tasks may include last-minute change notification using e-mail, twitter, instant messaging, or even an in-person message. Other options include the following:

- Intranet pages
- Blogs
- Wiki pages
- Verbal reports over tools such as VoIP or web conferencing

MANY-TO-MANY INTERACTIONS

In an online environment, many-to-many interactions, such as team meetings and class discussions, can be challenging. These work best if you act as a moderator, making sure that everyone contributes and is being heard. Because they are synchronous, chats can be particularly challenging. Distractions, poor typing skill, and limited bandwidth can also significantly impact how effectively students contribute.

ONE-TO-ONE INTERACTIONS

One-to-one interactions may include your virtual office hours, student reviews, or project-related topics, such as determining an approach with a particular student before bringing in the rest of the team. For these, you may choose to use phone calls or even face-to-face meetings, or virtual tools such as VoIP, videoconferencing, instant messages, or chat rooms.

GATHERING TEAM INPUT

In providing opportunities for many-to-one interactions, you might introduce brainstorming sessions at the beginning of a project, troubleshooting in the middle, or possibly a post-project review at the end. To gather team input, consider using a chat room or virtual meeting space; an e-mail list; a dedicated survey tool, such as Zoomerang or SurveyMonkey; or a survey on your team wiki.

CONCLUSION

Once you know the type of tool you will need, you can choose from many software packages available. Hundreds of tools exist for different types of collaboration, ranging from open source to highly specialized applications. Many tools offer more than one function. For example, with Windows Messenger, you can chat with one or more participants, hold voice and video conversations, share files, share applications, and save meeting notes on a whiteboard.

REFERENCES

Brown, K., Huettner, B., and James-Tanny, C. (2007) *Managing Virtual Teams: Getting the Most from Wikis, Blogs, and Other Collaborative Tools.* Texas: Wordware Press. The book's wiki is www.wikiwackyworld.com.

Wikipedia entry, online at http://en.wikipedia.org/wiki/Computer_supported_cooperative_work.

Bass, B. M. and Riggio, R. E. (2005) *Transformational Leadership* (*2nd ed.*). Mahwah, NJ: Lawrence Erlbaum Associates (10 Industrial Avenue) Associates.

Content Management Professionals (verified September 2006) http://www.cmpros.org/.

Duarte, D. L. and Snyder, N. T. (2001) *Mastering Virtual Teams: Strategies, Tools, and Techniques That Succeed* (*2nd ed.*). San Francisco: Jossey-Bass.

Ebersbach, A.,et al. (2006) *Wiki: Web Collaboration*. Berlin: Springer-Verlag.

Educause (2009) http://www.educause.edu/.

Kan, S. H. (2002) *Metrics and Models in Software Quality Engineering* (2nd ed.). Boston: Addison-Wesley.

Leuf, B. and Cunningham, W. (2001) *The Wiki Way*. Boston: Addison-Wesley.

Lipnack, J. and Stamps, J. (1997) *Virtual Teams: Reaching Across Space, Time and Organizations with Technology*. Wiley.

Mader, S. (2008) *Wikipatterns*. Hoboken, NJ: Wiley.

MetaCollab, an online collaboration about collaboration (http://collaboration.wikia.com/wiki/Main_Page).

Schafer, L. E. (2000) *How to Make Remote Teams Work*. Training materials from a seminar given to HP. Palo Alto, CA: Global Savvy.

Software Engineering Institute (verified September 2006) http://www.sei.cmu.edu/.

Software Project Survival Guide (viewed September 2006) http://www.construx.com/survivalguide/detailedchangeproc.htm.

University of Michigan School of Information Technology-mediated Collaboration Research (http://www.si.umich.edu/research/area.htm?AreaID=3).

Wikipedia (2006) *Change Control*. http://en.wikipedia.org/wiki/Change_Control (viewed September 2006).

Wilson, S. (2005) *InternetTeaming.com: Tools to Create High Performance Remote Teams*. Portland, OR: Inkwater Press.

CHAPTER

7

Communication Technologies

Anu Sivunen and Maarit Valo

Without the numerous tools and technologies now available to communicate and collaborate, virtual teams might never have become so widely successful. Because team members are often situated in distant geographical locations and since they rarely engage face-to-face, they must rely on robust communication technologies to accomplish their tasks.

Following their rapid development, many valuable tools are now available for use in virtual teams. Because of increased attention given to virtual teaming, scholars have recently become interested in exploring the use of these technologies, leading to a number of studies on their choice, use, and acceptance (see, e.g., Scott and Timmerman 1999; Venkatesh et al., 2003; Sivunen and Valo 2006).

Theories of Choice

How do practitioners choose the communication technologies they need? What do we know about the criteria they employ to

Virtual Teamwork: Mastering the Art and Practice of Online Learning and Corporate Collaboration. Edited by Robert Ubell

select tools they use for computer-mediated communication and virtual collaboration?

Scholars have been interested in these questions since the 1980s, at first employing terms such as "channel" and "medium." Today, "communication technology" and "information technology" frequently replace earlier terms. Still, the key question remains. On what basis do users choose among the various communication technologies available for different tasks and why do they prefer certain ones over others?

Practitioners select tools for either rational or social reasons (see Carlson and Davis, 1998; Fulk et al., 1990). *Rational choice* represents those based on rational criteria, that is, on the basis of which technology they deem most suitable for the task at hand. *Social choice* denotes those affected by, or dependent on, others.

Until now, conclusions have been drawn principally from studies of choices made in traditional organizations, such as large companies. Only recently have scholars taken an interest in choices made by those in virtual teams (see Scott and Timmerman, 1999; Sivunen and Valo, 2006; Sivunen, 2007). Nonetheless, we believe the rational vs. social dichotomy is applicable in both contexts. (Table 7.1 presents these theoretical perspectives and the principal theories they embrace.)

Rational choice acknowledges communication efficiency. When communication technologies are selected on the basis of

TABLE 7.1. THEORETICAL PERSPECTIVES OF COMMUNICATION TECHNOLOGY CHOICE

Rational Choice of Communication Technology ("Trait Theories Perspective")	Social Choice of Communication Technology
The social presence theory	The social influence model
The media richness theory	The symbolic interactionism
The access/quality approach	The theory of adaptive structuration

rational criteria, that is, on which one will suit the task best, communication tends to be most efficient (Fulk et al., 1990). In early research, rational choice was based on two well-known communication theories—*social presence* (Short et al., 1976) and *media richness* (Daft and Lengel, 1984). Social-presence theory believes that participants can achieve what they call "social presence." Proponents were highly sceptical of the capacity of computers to mediate social presence at all, concluding that the ideal medium is face-to-face contact. Consequently, investigators ranked media according to their ability to convey visual expressions, gestures, and vocal cues.

Based on a similar view, media richness theory described various media, including face-to-face communication, according to the "richness" of interaction they provided, that is, according to textual, vocal, and visual communication modes. Face-to-face interaction was rated as the richest, and consequently, other channels are poorer means of communication. The theory suggested that a lean medium, such as e-mail, is acceptable when messages are "simple," but when messages are equivocal, ambiguous, or emotional, or if nonverbal backchanneling cues or immediate verbal feedback are needed, richer media are required.

Another rational choice idea is known as the *access/quality approach* (Carlson and Davis, 1998). In this view, individuals select communication technologies on cost-benefit analysis, aiming to achieve a balance between how much effort is needed to access the medium and how acceptable the quality of information exchange is when using it.

Theories that favor rational choice can be gathered under "trait theories perspectives" in which the choice and the use of technology are explained by the characteristics or traits of the technology itself (Carlson and Davis, 1998). However, numerous studies have shown that the traits of technology neither determine nor predict user experience. Users are able to adapt to technologies and make the best of them once the choice has been made.

A radically different viewpoint is the social-choice perspective in which the social world, with its values, attitudes, and habits of other people, affects which medium is seen as best in a given situation, interpersonal relationship, or virtual team.

Social choice is based on three main theoretical viewpoints—*social influence model* (Fulk et al., 1987, 1990), *symbolic interactionism* (Trevino et al., 1987, 1990), and *theory of adaptive structuration* (Poole and DeSanctis, 1990). The social influence model emphasizes the changing character of our understanding of various media. It argues that our perceptions are not permanent or objective, but vary across contexts, situations, and tasks. It also recognizes the influence of coworkers on our choices.

Symbolic interactionism (originally presented by Mead, 1934, and Blumer, 1969) acknowledges the crucial importance of organizational culture. Although the choice of a certain tool for a given task may be made entirely by an individual, its origin is always socially constructed. Shared practices, symbols, and meanings are embedded in organizational culture.

A third model, *adaptive structuration*, occupies a position midway between the rational perspective and social perspective. Adaptive structuration emphasizes cultural and social factors, viewing choice as a product of particular cultures in which technologies are used. Organizational culture influences choice, and vice versa. Organizations and technologies also impact one another. According to this theory, use cannot be analyzed without also examining cultural factors. In practice, technology choice is seen as a collective decision.

Is it possible to claim that one or the other of these opposing perspectives is correct? Both have contribution to make. Theories of rational technology choice suggest that choice is based on rational thinking about the situation and the characteristics of the technology. It is easy to see the advantages of this principle. Shouldn't all important decisions be rational? However, the quality of team communication and team outcomes depend on a wide range of factors, of which technology is merely one. Communication efficiency does not increase by the "richness" of technology. Rather, efficiency arises from concentrating on what is essential. E-mailing, for example, or talking on the phone may be excellent choices when there is no need to acquire visual information; videoconferencing may be the best solution if it is critical to see facial expressions to arrive at a judgment.

Social technology choice recognizes the importance of social systems in the workplace. Virtual teams are social systems with their own communication history and culture as well as their own relationships, roles, and individual identities (see Chapter 2). These factors all come into play when decisions on which communication technology to use are made.

TECHNOLOGY CHOICE

Current empirical research on virtual teams and the use of communication technology is based largely on laboratory studies and zero-history student groups (see, e.g., review by Scott, 1999). There is a serious lack of empirical data on virtual teams. Findings presented here go some way toward rectifying what's missing in the literature. Data are drawn from a study of four virtual organizational teams (Sivunen, 2007), collected from three globally dispersed, cross-cultural teams and one nationally dispersed team. Members of global teams were located in nine different countries—Austria, Canada, Denmark, Finland, Germany, Great Britain, Norway, Sweden, and the United States. The national team consisted of members drawn from two cities in Finland. All team members and leaders ($N = 35$) were interviewed, and their actual, team-based communication using various communication technologies was observed and recorded. Tools most commonly used for communication were e-mail, telephone, an instant-messaging system, a discussion forum, and videoconferencing and call-conferencing systems. Each tool and its characteristics are analyzed and their functions in team communication discussed in the following text.

E-MAIL

In a study of technologies used in virtual teaming (Scott and Timmerman, 1999), e-mail and telephone showed the greatest usage. Even today, e-mail seems to be the most commonly used tool in virtual teams. The finding is rather surprising, considering the large number of other communication technologies for teamwork now on the market. Let's take a closer look at the reasons behind its success.

When team members in the four virtual teams under study were asked about technologies in team communication in general, the tool first mentioned was usually e-mail. There are many reasons for extensive use of e-mail in virtual teams. Even though usually there are no clear guidelines for e-mail, it turns out to be the most preferred tool, and its use results from its ubiquity in day-to-day routines.

Users engage in e-mail communication for a variety of reasons; chief among them are for giving information, for asynchronous communication, for storing messages, and for managing social distance.

In virtual teams, the use of e-mail as an information channel is one of its principal benefits. It is a good tool when you need to deliver a message to several recipients simultaneously and when the content is informative, rather than conversational.

E-mail is equally useful for asynchronous communication. When the subject is not urgent, e-mail is often the mode of choice because it gives the sender time to prepare a well-conceived message, and it also gives recipients time to reflect and write a well-thought-out response.

E-mail also has the decided advantage in that messages can be archived easily. With e-mail, both the sender and the recipient can access saved messages at will. Assignments are also easier to remember because they can be viewed quickly in in-boxes. As one virtual team member said, "You really cannot trust oral assignments. People forget them so easily. But if you write them through e-mail, they are remembered better. Then there is a note that something should be done."

Perhaps the most intriguing aspect of e-mail is its ability to manage social distance (see Chapter 2). Exploited one way, it can increase social distance; implemented in other ways, it can decrease social distance. In some situations, senders may intentionally take advantage of e-mail as a "lean" and asynchronous medium. By deliberately choosing e-mail, senders can give recipients time to think how best to express themselves. Alternatively, e-mail can also be used to reduce social distance. Despite its "leanness," it is possible to show caring and closeness. One virtual team leader said that she often chooses e-mail to make a friendly enquiry about remote team members. ". . . if I haven't heard from somebody for a long time," she said, "At least I have to send an e-mail and ask how s/he is doing."

Extensive use of e-mail presents certain disadvantages. Several team members felt that heavy use of e-mail can cause strain. Writing and reading e-mail can generate information overload, often taking time away from other tasks in order to read them and respond. In some teams, members sought to reduce use of e-mail and replace it with other communication media. The risk comes from sending e-mail blasts to great numbers in your network. Easily forwarded, messages can reach team members who have little or no direct interest in the information sent. The number of recipients may be many times greater with e-mail than with other tools. E-mail messages are also often sent and received across team and organizational boundaries, further increasing traffic.

In virtual teams, it is highly likely that e-mail will remain a central communication tool owing to its simplicity and its place in everyone's routine. Because of its ease of use and its asynchronicity, e-mail frequently triumphs over other vehicles that may deliver many more personal cues.

TELEPHONE, MOBILE PHONE, TEXTING

The use of the conventional telephone as well as mobile phones can be experienced quite differently across teams and among team members. In some virtual teams, phone calls and e-mails are used frequently, whereas in other teams, phones rarely ring. If team members are dispersed over several time zones, synchronous tools, such as the telephone, present serious scheduling obstacles. In most virtual teams, conventional telephones have been replaced by mobile phones, largely because participants are often on the move. Mobile phones offer the advantage of asynchronous messaging. With their text messaging facility, mobile phones have brought an asynchronous, text-based dimension to phone use (texting, SMS).

For members of virtual teams, telephones offer two key benefits—speed and synchronous communication. Since speed is often essential, with mobile phones, you can reach busy team members with text messages. Even when recipients are engaged in meetings—and if the matter is urgent—they can discreetly answer with a text message.

Synchronous communication by phone is often necessary when team members have something deeper to discuss with one another or when the issue is complex or ambiguous. Interviewed for this study, a team member from a global virtual team said, "If it's anything more specific on . . . an issue you want to discuss, then we would use the phone."

DISCUSSION FORUMS

Virtual teams typically use text-based discussion forums to support communication. In these forums, team members can share information and answer questions asynchronously. Recently, blogs and wikis have become popular, and in some organizations, they may be replacing more conventional discussion forums. While similar to the use of e-mail, text-based discussion forums, among other things, serve two critical functions—informing and message preservation—as well as asking and answering questions and decision-making.

In virtual teams, forums can be used equally for sharing information about small, team-related questions as well as for larger organizational discussions. Team leaders often exploit them as the principal channel to notify team members, leading to forums emerging as team bulletin boards. Leaders open them to share documents with team members. Discussion forums may also act as depositories where team-related messages and documents are stored, giving members an easily accessible place from which to download material. Among its most attractive features is the use of forums as a place where members post questions and where leaders can answer them broadly to all participants, rather than wasting responses useful to all in one-on-one e-mail.

Discussion forums can also act as decision-making centers where members offer their opinions asynchronously without overwhelming e-mail in-boxes. Members may also vote on team issues, allowing all participants to make their voices heard, especially when teams are globally dispersed. On occasion, polls may be employed to support decision-making at the organizational level. One virtual team leader who took part in the study encouraged members to participate in upper-level decision-making by asking their opinions in the discussion forum:

> Strengths and weaknesses at [Company X] websites?
>
> 12.03.2003–09:28:07
>
> Hello! We are doing a clearance [sic] to our management so that we can get a clear vision of our web development and our web strategy. One part of that is listing the strengths and weaknesses of our sites. So, give your opinion, what is good and what is not so good at our sites. Feel free to give your honest opinion.

INSTANT MESSAGING

Often used for one-to-one dialog, instant-messaging systems can now transmit both text chat and voice chat. To communicate with voice chat, the user needs to acquire a headset with a microphone and earphones. While many systems enable several participants to join a discussion simultaneously, most use it for dyadic communication. It is also common for participants in a text chat session to engage in several discussions at the same time, an effort that is far easier with messaging than in face-to-face settings.

Quite useful in virtual teamwork, instant messaging offers three principal benefits—asking and answering questions; making contact quickly; and relational communication. It makes it easy for team members to ask short questions since it provides simple and immediate access to team members who happen to be available, regardless of geography. Instant-messaging systems provide access to immediate user availability, giving them the means to display their status as "available," "busy," or "away," so that others know who can be contacted right away. It is often perceived as less interruptive than telephone calls.

Because messages automatically pop up on your screen, instant messaging attracts attention easily. If members are at a distant site and if they are online when you are, you can capture their attention and communicate on the spot. A virtual team member interviewed for the study remarked, "We have instant messaging on our computers and ... sometimes that's just much faster and more interactive ... when you see somebody is online, you can just send him a message knowing that it's gonna pop up on their screen right in there and that they will respond immediately."

Instant messaging can generate relational communication, allowing members to get to know one another, creating an *esprit de corps*, an electronic alternative to joining colleagues for coffee breaks, at the water cooler, or in the corridor. After discussing a task-related issue in an instant-messaging session, team members often share some relational communication as well. The following excerpt is taken from the end of an instant-messaging discussion between two distant team members:

A: You too, and enjoy your time off!!!:-))

B: Thanks, we are going to Sweden to celebrate our 1 year anniversary

A: congratulations to you both!!!

A: bye for now...

B: incredible it has been a year, so much has happened to me during the last year - a totally new life

B: Bye

A: me too during the past half a year...:-) but now I have to go - strategy session starts:-9

B: sounds exciting:-)

A: yep- but let's talk after your vacation:-) have a great time!!!

In some virtual teams, instant messaging may have overtaken telephone conversation. But it will not take hold as a medium unless there is a receptive culture in your organization. Employees at ease with traditional ways may find fast and informal instant messaging with its constant accessibility neither convenient nor comfortable. Nonetheless, the popularity of instant messaging is a sign that it may enter common practice soon.

Videoconferencing and Audio-Conferencing

Conferencing systems enable virtual team members to see or hear one another synchronously from a conference room or at their computers from distant sites. Some systems also enable file sharing or even allow users to share desktops and documents. Some systems are fairly generous in the number of participants they accommodate, sometimes giving several members communication rights. For video and conference calls with more than a handful of invited participants, it's best that a leader structures the event so that it runs smoothly. In virtual teams, conferencing systems allow members to share a common space, introduce newcomers, inform participants, engage in decision-making, participate in synchronous communication, and in videoconferencing only, permit the option of showing and pointing.

Sharing a common space is important for geographically dispersed virtual teams. With team members working in various locations, conferencing may provide them with a sense of sharing a common space, even though it's virtual. When members see or hear one another in similar offices, they may share a sense of copresence.

Introducing newcomers through conferencing gives established team members the chance to get to know the faces and voices of new colleagues. It's equally important for newcomers to be given a chance to engage with existing members so that they, too, can match faces and voices with names. Introducing oneself to others during an audio or videoconference and getting immediate feedback can make newcomers feel comfortable and welcome.

Informing plays a central role for many virtual teams. Videoconferencing and teleconferencing are held to update team members and increase awareness of projects and their status. One team leader always gave a quick update of each member's situation at the beginning of conference calls. "I thought we'd just start with a quick round of organizational issues," she said. "If there's anything specific going on at the moment in your countries, just share it with us, briefly, and then we'll go on to the normal agenda."

Decision-making is one of the key reasons to introduce conferencing. Despite geographical distance, they give members a chance to discuss issues and make decisions with every team member present. Even though the team leader may be responsible for making final decisions, conferencing gives members the chance to discuss possible conclusions before they are reached.

In contrast to e-mail, which is largely seen as inadequate at decision-making that requires participation of a number of key players simultaneously, synchronous conferencing allows members to discuss issues and solve them together at the same time.

Showing and pointing are considered critical when geographically dispersed participants need to examine images together. Usually, these operations call for pointing to something or showing something to illustrate options to others. A member of a virtual web development team said,

> Videoconferencing is very good if you want to look for example at what we have done with our [web]site, and if you want to discuss what you

> want to do with your site. Videoconference is very good because you can actually put it [picture of the site] on a screen and discuss it there, and both parties can actually see it. So I think the videoconference is a system to use if you want to physically look at something with more than one person.

Even though videoconferences are often preferred to teleconferences, the ability to see the other party does not necessarily give added value to communication. This finding is supported by a study that compares different features of videoconferencing and conference calls (Burgoon et al., 2002). If the showing and pointing functions are not an important part of team collaboration, teleconferencing may be sufficient and a more economic tool.

WEBCASTING

Virtual teams may also make use of webcasting, transmitting audio, video, and other image content over the Internet or in a local network using streaming video technology. Webcasts can be viewed synchronously and stored in the team's or organisation's archive and are available on demand. In virtual teams, webcasting may be especially useful in one-to-many presentations when a team leader or member gives a presentation to others or when your organization delivers a seminar or training session. Some systems enable receivers to submit questions while viewing the webcast, either in text or voice. Questions are stored in a central database and forwarded to the respondent. This feature provides virtual team members with a convenient way to participate.

RECOMMENDATIONS

There are several tools available that support communication in geographically distributed virtual teams. The nature of the social interaction made possible and achieved through these technologies may vary significantly, depending on your team's communicative goals, experiences, and skill. Team culture, the relationships between members and its leadership, may also have an impact on technologically mediated communication.

Communicative functions of various tools differ significantly from one another. Based on theoretical considerations and empirical results reported here, as summarized in Table 7.2, we propose a set of applicable tools for each communication function in virtual teams.

Links shown between functions and tools in Table 7.2 are far from exhaustive since the goals and modes of communication in teams are unlimited. However, the table gives some insight into team communication. One tool is seldom enough for virtual teaming.

Empirical findings on the use of communication technology in cross-cultural virtual teams (Sivunen, 2007) show that team leaders and members choose and use communication tools rather randomly. There may be several reasons for this.

First, instead of rationally selecting from various options, choice seems to be based on team leader or member habitual preferences. Team leaders may introduce tools that they may already be familiar with, unaware of the wide range of others available. Choices may also be highly individual, so that even team leaders themselves are unaware of tools employed by their own team members.

Second, virtual teams usually fail to take deliberate steps to agree on or develop guidelines for communication technologies. Consequently, we strongly recommend that team leaders encourage team discussion on the choice and use of communication tools at an early stage in the life of the team. Agreement should be based on knowledge of the characteristics of both the requirements of team communication and the technologies available to satisfy them.

TABLE 7.2. THE MOST IMPORTANT COMMUNICATIVE FUNCTIONS IN VIRTUAL TEAMING AND THE COMMUNICATION TOOLS APPLICABLE IN EACH CASE

Communicative Function	Communication Tool
Informing or updating one team member	E-mail
	Text message
	Instant messaging
	Telephone
Informing or updating several team members	E-mail
	Text message
	Instant messaging
	Discussion forum
	Call conference
	Videoconference
	Webcast
Asking and answering urgent questions	E-mail
	Telephone
	Text message
	Instant messaging
Brainstorming, sharing ideas	Discussion forum
	Call conference
	Videoconference
Problem solving between two team members	Telephone
	Instant messaging
Problem solving in the team	Call conference
	Videoconference
Decision-making between two team members	Telephone
	E-mail
Decision-making in the team	Discussion forum
	Call conference
	Videoconference
Negotiating	Telephone
	Call conference
	Videoconference
Giving synchronous feedback	Telephone
	Instant messaging
	Call conference
	Videoconference

TABLE 7.2. (CONTINUED)

Communicative Function	Communication Tool
Getting in touch quickly	Telephone
	Text message
	Instant messaging
Relational communication	Instant messaging
	E-mail
	Telephone
	Text message
	Discussion forum
	Call conference
	Videoconference
Communicating asynchronously	E-mail
	Discussion forum
Communicating synchronously	Telephone
	Text message
	Instant messaging
	Call conference
	Videoconference
Storing messages	E-mail
	Text message
	Discussion forum
Visual back-channeling	Videoconference
Audio back-channeling	Phone
	Call conference
Creating social distance	E-mail
Creating social proximity	Instant messaging
	Telephone
	Call conference
	Videoconference
Sharing a common space	Call conference
	Videoconference
	Discussion forum
Sharing and discussing documents	Discussion forum
	Call conference
	Videoconference
Presenting, showing, and pointing	Videoconference
	Webcast

Based on theory and our empirical findings, we present a set of guidelines to help virtual team leaders select appropriate communication tools, helping to analyze criteria in selecting technologies and identifying best practices. These proposals may also help when establishing a new virtual team and deciding which tools best support the communication needs of team members.

GUIDELINES FOR SELECTING THE RIGHT COMMUNICATION TECHNOLOGIES FOR YOUR VIRTUAL TEAM

Tip 1. Seriously consider the communication network for your team. You will probably need to introduce both one-to-one as well as one-to-many communication.

Tip 2. Decide which communicative functions are most important for your team (see Table 7.2).

Tip 3. Consider whether there may be a need for constant availability, or whether it is better if team members can be reached only at certain times and in restricted ways.

Tip 4. Ask team members about their preferences and routines, but do not select technologies solely on the basis of their responses.

Tip 5. Be open-minded. Do not content yourself with already available technologies.

Tip 6. Review new communication tools, considering both their technical characteristics and the communication potential they offer your team.

Tip 7. Do not compare virtual technology with face-to-face communication as face-to-face is not always best. Communication tools enable your team to work. Without the tools, there would be no virtual team at all.

Tip 8. Technologies have their pros and cons, and all of them make relational communication possible.

Tip 9. Consider whether your team will need vocal, audio, and visual communication channels.

Tip 10. Consider your own as well as your team members' routine technology use and technologically mediated communication.

Tip 11. Discuss the technology you use commonly with your team members. Evaluate the technology itself as well as your own technologically mediated communication.

Tip 12. Tools available on the market are not necessarily better than open-source tools available free of charge.

Tip 13. Acquaint yourself with security and privacy matters.

Tip 14. Avoid introducing significant differences between your team's communication tools and those in use for the entire organization.

Decisions about communication technology in virtual teams are complex and need careful consideration by team leaders and members. In principle, the choice of technology should be as rational as any other team activity. Knowledge of the characteristics and possibilities of various communication technologies enables team leaders to provide exactly the tools that meet the needs of their team. However, communication technology use is always part of the social dynamics of the team, and an effective leader must also be conscious of the social side of technology use. It pays to make the best possible use of currently available tools and search for new solutions in response to team needs. A wise team leader also lets the team adapt communication technologies to its social structure.

It is important to remember that the right choice of a particular tool is only one part of the process of a successful technologically mediated team. The way people perceive technologically mediated communication in general may also be related to how communication functions through these tools. Sivunen (2007) found that virtual team members perceive technologically mediated communication in many ways. These perceptions may guide the selection of communication tools. However, it was found that when perceptions were compared with actual behavior, the two differed significantly from one another. In other words, team members are not always aware of the precise ways in which they communicate, often leading to inappropriate decisions.

Some of the criteria presented here for choosing and using communication technology may be more important in some teams than in others. Communication technology is being developed all the time, and future virtual team leaders should look for new possibilities and new features to support communication needs. Developers of tools need to be ready to meet the challenges that arise from user preferences. Since virtual team communication needs are likely to remain more or less the same in the short run, the most important challenges are to design tools that are both convenient and easily accessible everywhere.

The findings reported in this chapter are based on the first author's doctoral dissertation, supervised by the second author.

REFERENCES

Blumer, H. (1969) *Symbolic Interactionism: Perspective and Method*. Englewood Cliffs: Prentice-Hall.

Burgoon, J. K., Bonito, J. A., Ramirez, Jr.A., et al. (2002) Testing the interactivity principle: effects of mediation, propinquity, and verbal and nonverbal modalities in interpersonal interaction. *Journal of Communication*, 52(3), 657–677.

Carlson, P. J. and Davis, G. B. (1998) An investigation of media selection among directors and managers: from "self" to "other" orientation," *MIS Quarterly*, 22(3), 335–362.

Daft, R. L. and Lengel, R. H. (1984) Information richness: a new approach to managerial information processing and organization design. In: Cummings, L. L. and Staw, B. M., editors. *Research in Organizational Behavior*. Greenwich: JAI Press, vol. 6, pp. 191–234.

Fulk, J., Schmitz, J., and Steinfeld, C. W. (1990) A social influence model of technology use. In: Fulk, J. and Steinfeld, C. W., editors. *Organizations and Communication Technology*. Newbury Park, CA: Sage, pp. 117–140.

Fulk, J., Steinfeld, C. W., Schmitz, J., and Power, J. G. (1987) A social information processing model of media use in organizations. *Communication Research*, 14(5), 529–552.

Mead, G. H. (1934) *Mind, Self and Society*. Chicago: University of Chicago Press.

Poole, M. S. and DeSanctis, G. (1990) Understanding the use of group decision support systems: the theory of adaptive structuration. In: Fulk, J. and Steinfeld, C. W.,editors. *Organizations and Communication Technology*. Newbury Park, CA: Sage, pp. 173–193.

Scott, C. R. (1999) Communication technology and group communication. In: Frey, L. R., Gouran, D. S., and Poole, M. S.,editors. *The Handbook of Group Communication Theory and Research*. Thousand Oaks, CA: Sage, pp. 432–472.

Scott, C. R. and Timmerman, C. E. (1999) Communication technology use and multiple workplace identifications among organizational teleworkers with varied degrees of virtuality. *IEEE Transactions on Professional Communication*, 42(4), 240–260.

Short, J., Williams, E., and Christie, B. (1976) *The Social Psychology of Telecommunications*. London: John Wiley.

Sivunen, A. (2007) Vuorovaikutus, viestintäteknologia ja identifioituminen hajautetuissa tiimeissä (Social Interaction, Communication Technology and Identification in Virtual Teams). Dissertation. Jyväskylä: University of Jyväskylä.

Sivunen, A. and Valo, M. (2006) Team leaders' technology choice in virtual teams. *IEEE Transactions on Professional Communication*, 49(1), 57–68.

Trevino, L. K., Daft, R. L., and Lengel, R. H. (1990) Understanding managers' media choices: a symbolic interactionist perspective. In: Fulk, J. and Steinfeld, C. W.,editors. *Organizations and Communication Technology*. Newbury Park, CA: Sage, pp. 71–94.

Trevino, L. K., Lengel, R. H., and Daft, R. L. (1987) Media symbolism, media richness, and media choice in organizations: a symbolic interactionist perspective. *Communication Research*, 14(5), 553–574.

Venkatesh, V., Morris, M. G., Davis, G. B., and Davis, F. D. (2003) User acceptance of information technology: toward a unified view. *MIS Quarterly*, 27(3), 425–478.

PART

3

ENTERPRISE AND GLOBAL TEAMS

CHAPTER

8

TEAMING ACROSS BORDERS

RICHARD DOOL

Sergei, Li, Nathanial, Joanne, Tom, Adriana, Will, Ivonne, Christophe, Mike, Crista, Harpreet, and Jennifer all are members of an online graduate class at a university in New Jersey. They are all seeking graduate degrees in communication. Ten years ago, it would have been highly unlikely that they would be sharing this educational experience. Owing to distance, cultural barriers, technology, and limited expectations, a decade ago, very likely, they would have been in a more homogeneous class on campus. Today, all that has changed. More than 2.7 million students are pursuing education transnationally, with the United States leading the way, with more than 570,000 foreign students. In 2010, more than 4 million students will be learning virtually.

Sergei is from Yugoslavia, Li from Taiwan, Nathanial from Nigeria, and Joanne from New Jersey, as is Tom, Mike, and Jennifer. Adriana is from Romania. Ivonne is from New Jersey, a recent immigrant from Colombia. Christophe is from Greece originally. Crista is also from New Jersey, by way of Costa Rica. Harpreet is from India, currently living in London. Technology and globalization have largely reduced barriers to education. A

Virtual Teamwork: Mastering the Art and Practice of Online Learning and Corporate Collaboration. Edited by Robert Ubell

new paradigm is emerging, the "global classroom." As in the workplace, global classrooms have led to increased numbers of multicultural teams, online and on site. Teaming across borders refers to heterogeneous, multicultural teams located across geographic borders or multicultural teams all located within the United States. I refer to multicultural across borders as MAB teams and multicultural within borders as MWB teams. The distinction is not all that significant in terms of how to organize or manage an online class, but it does impact how instructors interact with online students in virtual teams.

TRENDS BEYOND AND IN THE CLASSROOM

Organizations today face a macro-environment with an unprecedented level of active "stressors"—technological advances, increased globalization, a nomadic workforce, economic shifts, heightened competition, rapid pace of industrial change, and increased diversity (Hamel, 1998; Beinhocker, 1999; Voelpel, 2003) (see Figure 8.1). Rapid change can generate

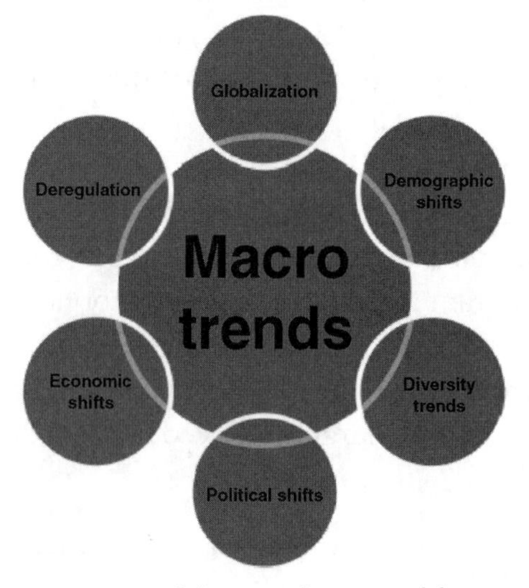

FIGURE 8.1. Macro-environmental forces.

continuous feelings of being rushed and of having to be "on" for long periods.

"Temporal windows of downtime between acute stressor events appear to be shrinking and are placed in a context of chronic pressures to learn and adjust to the day to day demands of technology and competitiveness in an increasingly global marketplace" (Sikora et al., 2004, p. 4). While many challenges faced today by organizational leaders are the same as those in the past, the pace and complexity of change is of a magnitude never before experienced (Beckhard and Pritchard, 1992). Lombardi (1996) noted that "constituent demands are escalating ever faster and more articulately" (p. xi). Many of the stressors occur simultaneously or overlap significantly. These trends—particularly those related to globalization, technology, and demography—have a direct impact on virtual teams as well as the traditional college classroom.

Technology has reduced spatial and temporal barriers and created increased opportunities for interaction. We can now connect seamlessly with customers, suppliers, employees, and others, no matter where they are located. We can employ workers beyond traditional commuting distances and deploy resources as needed around the globe. Technology has enabled the global classroom by reducing traditional barriers, linking widely separated students in a common educational experience.

Demographic trends have also had an impact. While white Americans are still the dominant source of students in the United States, the nation's higher education is becoming increasingly diverse. It is expected that almost 40% of the US workforce will be comprised of non-white workers within 20 years.

Much of the growth will come from China, India, and sub-Saharan Africa (see Figure 8.2). Hispanics in the United States are the fastest growing population segment, with Asians the second (see Figure 8.3).

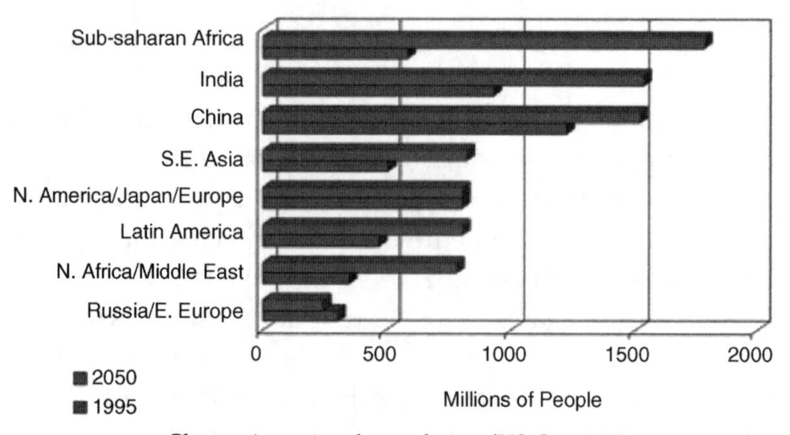

FIGURE 8.2. Change in regional population (US Census Bureau, 2000).

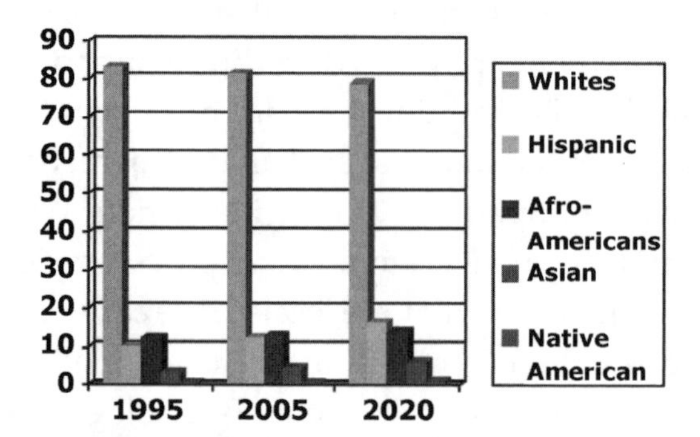

FIGURE 8.3. US population projections (US Census Bureau, 2000).

A Culture Primer

In a survey of the literature, Dadfar (2001) concluded that the nation would confront a number of multidimensional workplace cultural changes. Many of these contradictory phenomena are likely to be found not only in the global campus, but also in virtual teams—individualism and collectivism (Hofstede), polychronic time versus monochronic time (Hall), high context versus low context (Hall), affective versus neutral, and achievement versus ascription (Trompenaars).

INDIVIDUALISM VERSUS COLLECTIVISM

In highly individualistic cultures, people are expected to be self-reliant and independent, focusing on caring for themselves and their families, and remaining emotionally independent of particular social, organizational, or institutional affiliations. When they speak, individualists usually emphasize "I," not "we." In contrast, in highly collectivist cultures, people are expected to serve the group to which they belong, encouraging collaboration with others. When speaking, collectivists commonly emphasize "we," rather than "I."

Students from collectivist cultures tend to be more supportive of teamwork, preferring rewards that provide group incentives. They often seek group recognition over personal reward. Some may even view recognition that singles them out as a disincentive to achievement. Students from individualistic cultures, on the other hand, often prefer individual assignments. In the United States, for example, many students resist group assignments, troubled that their grades will depend heavily on the work of others. They tend to avoid teamwork, fearing that it will not contribute usefully to their personal goals. The United States, Australia, and Great Britain are examples of highly individualist cultures. Most Asian and Latin American countries are highly collectivist (see Figure 8.4).

HIGH CONTEXT VERSUS LOW CONTEXT

High-context versus low-context cultural dimension refers to communication norms. In low-context countries, communica-

	Coutries
Highly Individualistic	United States
	Australia
	Great Britain
	Canada
	New Zealand
	Italy, France, Ireland, Germany
	India, Japan
	Mexico
	China
Highly Collectivist	Latin American Countries

FIGURE 8.4. Individualistic and collectivist countries. Adapted from Dadfar (2001).

tion relies significantly on the literal meanings of words expressed. In these cultures, written and spoken communication tends to be very explicit. Low-context individuals tend to be verbally explicit, precise, and accurate. They do not assume that others will be able to figure out what they mean without a great deal of help.

Bluntness and directness are commonly expressed by low-context speakers. To others in high-context cultures, the style may seem insulting or aggressive. For people who come from high-context cultures, meaning is often derived more from the context in which written and spoken communication is embedded. When communicating with others, high-context cultural groups tend to be more verbally implicit, expressing themselves more indirectly and subtly, relying frequently on nonverbal cues. High-context cultures often depend more on what is *not* said, than on what is actually expressed. Rather than propose ways of thinking or acting explicitly, high-context speakers often use suggestions or offer alternatives.

In high-context cultures, the social status of who is speaking and who is receiving the message, as well as the nature of the relationship between them, strongly affects how they communicate with one another. The meaning of everything said in high-context cultures must be interpreted in the context of the social relationship between individuals. Germany, Switzerland, and the United States are examples of low-context countries. China, Korea, Japan, and most Latin American countries are examples of high-context countries (see Figure 8.5).

MONOCHRONIC TIME VERSUS POLYCHRONIC TIME

In cultures where a monochronic view of time is prevalent, people tend to do only one activity at a time, keep a strict schedule of their appointments, and show a strong resistance to

Countries	Countries
High Context	Japan
	China
	Italy, Spain, France
	Latin America Countries
	Middle Eastern Countries
	India
	Koreas
	British
	United States
	Germany
Low Context	Austria

FIGURE 8.5. High- and low-context countries. Adapted from Dadfar (2001).

deviating from plans. In cultures where a polychronic view of time is the norm, people tend to do more than one activity at a time, appointments are approximate and may be changed at any time, and schedules are not as important as relationships. Inflexible adherence to schedules and plans is only as beneficial as the quality of those schedules and plans. North Americans and Northern Europeans tend to have a monochronic view of time. Mediterranean, Latin American, and Arab cultures tend to have a polychronic view (see Figure 8.6).

AFFECTIVE VERSUS NEUTRAL

In highly affective cultures, people tend to openly express their feelings. In highly neutral cultures, emotions are not expressed as openly and naturally. People from highly affective cultures are more likely to smile, talk loudly when excited, and greet each other enthusiastically. People from highly neutral cultures experience the same emotions, but are less inclined to express them, and they express them more subtly. Implications for behavior in the classroom include how demonstrative people are when showing appreciation and affection for each other.

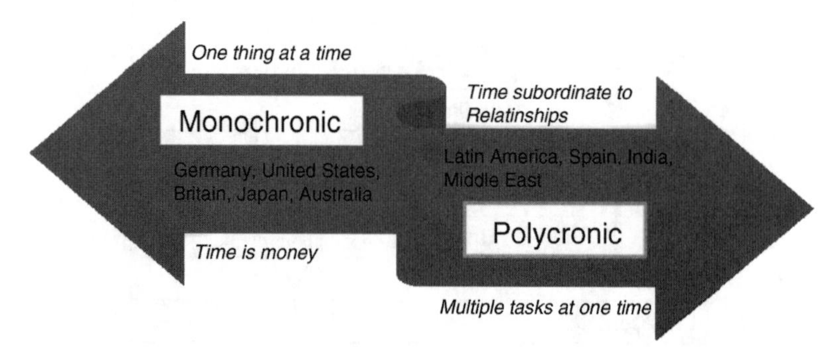

FIGURE 8.6. Monochronic and polychronic perceptions of time.

Mexico, the Netherlands, Switzerland, and China are examples of highly affective cultures. Japan, Britain, and Singapore are examples of highly neutral cultures. The United States is in the middle."

Achievement versus Ascription

An achievement orientation judges people on what they have accomplished and on their record of results (achieved status). Ascription refers to the status that is attributed to one through birth, kinship, gender, age, interpersonal connections, or educational institution (ascribed status). Achieved status refers to accomplishment; ascribed status refers to belonging.

In highly achievement-oriented cultures, social status is largely derived from a person's achievements. In highly ascription-oriented cultures, social status is largely derived from personal attributes such as age, experience, social connections, or gender. In organizations, a person's status is reflected in his or her privileges such as access to resources and perks, deferential treatment, and input in decision-making. Australia, the US, Switzerland and Britain are examples of highly achievement-oriented cultures. Venezuela, Indonesia and China are examples of highly ascription-oriented cultures.

Consequently, it is important for instructors to understand different orientations that students exhibit when they come from achievement or ascriptive cultures or high- and low-context cultures. While these considerations are not precise or exhaustive, they offer some insight about common student differences that may affect participation in discussions, expectations about the course, student-to-student interaction, as well as student–instructor interaction. (Table 8.1 offers an overview of the major cultural differences.)

TABLE 8.1. SUMMARY OF HIGH- AND LOW-CONTEXT MACRO-DIFFERENCES

Context	Personal Relationships	Time	Formality	Affective/ Neutral	Achievement/ Ascription
High context	Very important, more than details	Relationships trump punctuality. Honors "events" over deadlines	High degree of formality	More affective	More ascriptive
Low context	Unimportant compared to grade and details of assignment	Punctuality is valued, linear time orientation	More informal	More neutral	More achievement oriented

CASE IN POINT

CULTURAL DIFFERENCES EXHIBITED BY US STUDENTS AND TAIWAN STUDENTS

According to Hofstede, the United States is an individualistic culture, and Taiwan a collectivist. In the United States, achievement and competition are reinforced continuously, while in Taiwan the norm encourages cohesiveness and family. The United States is less accepting of status differences than Taiwan (see Table 8.2 for classroom implications).

ONLINE STUDENTS

In its Fifth Annual Report, *Five Years of Growth in Online Learning*, the Sloan Consortium (2008) reports that:

> The number of students taking at least one online course continues to expand at a rate far in excess of the growth of overall higher education enrollments. The most recent estimate, for fall 2007, places this number at 3.94 million online students, an increase of 12.9 percent over fall 2006. The number of online students has more than doubled in the five years since the first Sloan survey on online learning. The growth from 1.6 million students taking at least one online course in fall 2002 to the 3.94 million for fall 2007 represents a compound annual growth rate of 19.7 percent.

MORE FOREIGN STUDENTS

The United States leads the world in foreign students with 28% of the market. Britain is second at 12%, followed by France (11%) and Germany (10%). Foreign students spent more than $13 billion in the United States (Hira, 2003). Chow and Marcus (2008), of the Institute of International Education, note that:

> In 2005, more than 2.7 million students were pursuing transnational higher education—a 47 percent increase over the 2000 figure of 1.7 million students. A concurrent increase has occurred in the number of

TABLE 8.2. SUMMARY OF US AND TAIWAN STUDENT DIFFERENCES

	US Student	Taiwanese Student
Group activities	Prefers individual work	Will work harder in a team activity due to collectivist orientation
Class requirements	Prefers flexibility in class requirements (attendance, participation, due dates)	Expects mandatory requirements due to culturally embedded respect for authority
Informality	Expects access to the instructor outside of class and a certain level of informality	Expects a higher level of formality. Would hesitate to contact instructor "outside" of class
Grading concerns	Is more likely to express grade concerns or to appeal a grade	Is unlikely to question or appeal a grade
Debating with peers and the instructor	More comfortable debating with the instructor in online discussions	Will be unlikely to question or debate an instructor during online discussions
Content expectations	More accepting of practical applications of course materials	Will be more accepting of theoretical, research-based material

Source: Niehoff et al. (2001).

TABLE 8.3. FOREIGN STUDENTS STUDYING IN THE UNITED STATES (2006-2007)

Foreign Students in the United States (2006–2007 School Year)	Country of Origin	Students in the United States
573,000 (4%)	India	83,800
	China	67,700
	South Korea	62,300
	Japan	35,200
	Middle East	22,300
Graduates		45%
Undergraduates		41%
Certificates, nondegree		14%

students seeking an international education in nontraditional destinations in Asia, Africa, and Latin America. Several countries in these regions have positioned themselves as key actors in the global economy, attracting more students to their shores. Despite these developments, the United States continues to be the top host country for students seeking higher education abroad. In 2006, the United States attracted 30 percent of internationally mobile students among the leading eight host countries (Australia, Canada, China, France, Germany, Japan, and the United Kingdom).

In 2006–2007, a total of 573,000 foreign students studied in the United States (see Table 8.3), an increase of 3% over the prior school year; 59% of these students come from Asia, with India and China leading the way with 10% and 8% growth over the prior school year. The United States also saw a 25% increase in foreign students from the Middle East.

MULTICULTURAL VIRTUAL TEAMS

The introduction of virtual teams in industry is growing rapidly, driven primary by the ability to gather workers seamlessly from disperse locations. Despite geographic dispersion, virtual teaming gives organizations the ability to tap into expertise, experience, and capabilities for projects anywhere in the world. Virtual teams can help disperse improved processes across an organization, support cross-functional and cross-divisional interaction, save time and money by avoiding travel expense, and improve the quest for talent by eliminating artificial geographic barriers (Johnson et al., 2001).

Because of globalization and demographic changes, the multicultural virtual team (MVT) has fully emerged. In many organizations and in the classroom, virtual team membership often crosses national boundaries, with many cultural backgrounds represented. Research has found that MVTs, after a period of adjustment, outperform homogenous teams, especially in problem solving and creative thinking (Taras and Rowney, in pressTaras and Rowney, n.d.). MVTs are also less likely to be affected by groupthink, stimulated by divergent insight and perspectives.

Still, MVTs face challenges. People from different cultures vary in their values, personalities, and work and communication styles (Shin, 2005). Often, these differences—combined with spatial, temporal, and technology issues—can cause conflict and undermine trust. Owing to geographic and time differences, virtual teams, and MVTs in particular, can experience task, role, and responsibility ambiguity, creating obstacles to success.

"Global Classroom"

Clearly, the introduction of teams in industry and education is unlikely to decline. CEOs routinely list "teamwork" among the most highly prized attributes of new employees. Given current demographic and globalization trends, the ability to contribute positively in a multicultural team will be a favored career skill. Instructors have an obligation to help students learn how to be effective members of multicultural teams, especially in MVTs.

Unfortunately, MVTs create opportunities for cross-cultural misperception, misinterpretation, and misevaluation (Adler, 2008). Misperception is commonly founded on learned and culturally determined ideas about others. We "see" and anticipate behavior in others based on our own cultural filters. In MVTs, each student's perceptual filters are different, affecting how each one is able to "team," and how they interpret priorities, roles, and course expectations. Cross-cultural interpretations are based on how an individual assigns meaning to their experiences, observations, and relationships. Assumptions play a significant role. Misinterpretations come from a variety of sources—stereotypes, subconscious cultural blunders, parochialism, and a lack of cultural self-awareness (Adler, 2008). Cultural misevaluations may emerge from cultural conditioning. Most of us tend to view others using our own cultural vision as the standard, leading to common miscalculations and misinterpretations.

CHALLENGES

MVTs in online courses can be accompanied by an array of potential challenges, relating to language, time, ethical values, expectations, perception, and attitude. All can be managed successfully if understood and anticipated.

LANGUAGE

By definition, online courses tend to be based on the written word, exploiting text-based discussion, posted course documents, written assignments, and e-communication. Of course, visual elements are often added, as well as real-time interactions—conference calls or synchronous group meetings, for example. Even with more advanced applications, such as podcasts, online courses are likely to be predominantly text-based. Across the myriad of interactive possibilities, language is central. Your MVT is likely to encounter a wide range of English capabilities. Some team members will be native English speakers; others may understand English as a second or third language. Of course, there are also wide differences in what constitutes "English" across regions and national borders, particularly in expressions or colloquialisms. Nonnative English speakers may not appreciate the language to the same degree as native speakers, and, as a result, may be reluctant to express themselves freely. They also may have trouble expressing themselves clearly that may lead to miscommunication, misunderstandings, and misinterpretation (Payne, 2008).

Time

As MVTs cross borders, online instructors may encounter differences related to monochronic and polychronic time orientations. Students from a monochronic culture are apt to focus on deadlines, due dates, priorities, and a linear progression of assignments. Polychronic students often view deadlines and priorities as more fluid, tending to subordinate schedules to relationships within the team. Likely to be better at multitasking, they tend to not get as caught up in details as their monochronic virtual team members.

Ethical Values

To a large extent, ethics is culturally bound. Ethical standards differ by culture, country, and region, and even individually. Ethical norms also emerge from various institutions students may come in contact with—employers and universities, for example. While many ethical considerations are broad, in an MVT course, the key issue most likely to confront students and instructors is plagiarism. Plagiarism is a major source of online conflict.

There are two principal causes of ethical questions in MVTs. First, there are clear differences among cultures relating to achievement, time, and relationships—areas fraught with opportunities for ethical tension. The second, far more practical, concerns standards. Most US colleges and universities use guides prepared by the APA or MLA as their academic standard. Instructors should not assume that international students in their classes understand these standards and can apply them appropriately.

CASE STUDY

"Sharmistra" (from India) was a graduate assistant in a master's program. The director of graduate studies was surprised and disappointed when he received a call from one of the department professors informing him that "Sharmistra" had plagiarized material on a recent paper. It had come back with a rainbow of colors on her Turnitin report. The director now faced the dilemma of supporting the professor, enforcing the department and university ethical code of conduct, and managing the graduate assistant, who now was at risk of losing her graduate assistant position, stipend, and academic reputation.

The director read the submitted paper, the course syllabus, and the Turnitin report. "Sharmistra" had clearly used many sources for her paper and worked diligently on the assignment. Her writing style was consistent with other papers she had submitted and none came back with any concerns after being submitted to Turnitin as a result of this accusation.

The director met with the student to discuss the situation and its ramifications. It took awhile for "Sharmistra" to grasp what she had done wrong as well as the potential risks. It became very clear that student did not understand the notion of "paraphrasing" and the rights attached to electronic sources as they relate to APA standards (in use at the university). School and department documents had addressed these issues, but on review, it became clear that a series of assumptions were made in preparing them. For example, it was assumed that students understood the notion of paraphrasing and instructions were not as detailed as they could have been. Documents were all written as part of a US cultural paradigm.

As a result, resources were rewritten against a broader multicultural framework, specific areas in question were more fully explained, and a seminar on APA was organized for all international students when they enter the university.

"Sharmistra" was put on warning, given a failing grade for the specific assignment, and retained her graduate assistantship on a probationary basis. While there were no further infractions, some quite positive changes emerged.

Expectations, Perception, and Attitudes

Students who enroll in online courses and who join MVTs do so with varying expectations, perceptions, and attitudes, creating challenges for instructors as well as for themselves. It all starts with expectations (Niehoff et al., 2001). Students in culturally diverse classes hold different expectations about their courses and their teams. If expectations are unmet, it can affect student learning and team performance, and may lead to withdrawal and absence or "going silent." Research shows that expectations of international students are rarely considered in the US-based course practice and administration (Niehoff et al., 2001).

There are ample opportunities for misperception. Teammates may inappropriately stereotype colleagues, unconsciously assuming that national stereotypes apply to members in their team and, as a result, participants may judge colleagues inappropriately (Adler, 2008). "Members of culturally diverse teams express higher levels of distrust than do their more homogenous counterparts" (Adler, 2008, p. 135). Students may be more in tune with members of their own culture, or even their own region, threatening productive discussions and teaming.

MAB AND MWB TEAMS

As noted earlier, MAB teams are online student teams comprised of members who live and work in different countries. For example, MAB teams may draw students from the United States, United Kingdom, Nigeria, and Taiwan. MWB teams, on the other hand, are online student teams comprised of members from various cultures but who are all physically located in the same country.

Instructors are likely to encounter both kinds in their online courses as more programs go fully online and as increasing numbers of students seek education abroad. Since the principal difference between the two is instructor attitude and assumptions, it's best if both are taught by culturally aware faculty. Unfortunately, many MWB courses tend to be nation-centric, failing to recognize cultural diversity among students. Instructors often assume that since students are all physically present in one country, common cultural norms are practiced by all, a conclusion that ignores high- and low-context variances among students as well as the effects of different time orientation, communication style, teaming, and ways of resolving conflict.

Instructor as Multiculturalist

In MAB and MWT courses, the instructor takes on an added role as "multiculturalist," a quality particularly important in the design and management of MWTs. As multiculturalists, instructors must learn to be aware of their own cultural biases, examining their own beliefs, values, behaviors, and prejudices (Ford, 2001). They must also be aware of the social, political, and economic context of the course as well as what is occurring in the students' home country.

As multiculturalists, instructors must embrace difference, rejecting the notion that students are in a "melting pot," asked to blend into a single pattern of expectations. Honoring difference, instructors must create an atmosphere in which individual cultures enhance the class and the team, fostering a global attitude from which students emerge with broadened perspectives.

Culturally aware faculties are more open, ask more questions, and listen more carefully to their students. At their best, they create a culturally neutral environment, allowing students to interact as seamlessly as possible. Multiculturalists treat their students "as they would like to be treated" (Ford, 2001, p. 4). Multicultural faculties assume that online communication is all cross-cultural.

Multicultural Class Components

To create a positive learning experience in your multicultural course and to enhance teaming, it's useful to offer a set of components that consider diversity and embrace difference as an educational asset. It's prudent to introduce elements that provide clear expectations, an active socialization–orientation process, overlapping communications, culturally neutral language, expanded feedback, and a sense of community (see Table 8.4).

TABLE 8.4. MULTICULTURAL CLASS COMPONENT CHECKLIST

Multiculturalism Class Component	Checklist
Creating a sense of community	☑
Clear expectations	☑
Socialization/orientation process	☑
Overlapping communication	☑
Clear, unambiguous language	☑
Appropriate and timely feedback	☑

CREATING A SENSE OF COMMUNITY

In multicultural classes, it's essential that instructors embrace a sense of community. In online courses, community can open communication, collaboration, interaction, and participation, building connection and belonging, especially for those from high-context cultures (Grzeda et al., 2008). Community can be created with a series of instructor actions (see Table 8.5):

- Post an open, positive "welcome" message that celebrates multiculturalism, setting the stage by promoting the idea that difference should be embraced and celebrated, citing the value of effective virtual teaming as a career competency.

TABLE 8.5. CREATING A SENSE OF COMMUNITY CHECKLIST

Creating a Sense of Community	Checklist
Welcome announcement	☑
Personal welcome e-mails to each student	☑
"Live" interaction if possible (welcome conference call)	☑
"Soft touch" in managing course boundaries and enforcing course expectations	☑
Consistent reminders	☑
Praise in public, critique in private	☑
Consistent inclusionary approach	☑

- Send personal e-mail welcome messages to each student, proposing that they introduce themselves, giving them an opportunity to share their experiences.
- If possible, arrange a live interaction with each student. In some schools, instructors are encouraged to schedule a real-time welcome call with each student. While this approach presents logistical challenges, nonetheless, it can be very effective, especially with team members from abroad.
- To enforce your course expectations as well as suggest changes in student behavior, it's best to employ a "soft touch," expressing yourself sensitively and positively.
- Introduce consistent reminders of the value of cross-cultural interactions.
- Praise positive teaming in public and offer critiques in private.
- Offer a consistent, inclusive approach in all interactions and communications.

Clear Expectations

Creating clear expectations is especially essential in online multicultural classes and MVTs, given varying student expectations. With distinct differences in context, time orientation, communication styles, and formality, it is essential that your expectations are delivered clearly.

Socialization and Orientation

Acculturation occurs more quickly if members of MVTs are given a chance to get together before teaming activities begin. Logistically challenging, face-to-face opportunities are often unavailable to online students, especially since time-zone and geographic obstacles interfere. Despite hurdles, there are several options at your disposal that you can employ to encourage acculturation and socialization, critical elements in generating productive teaming (see Table 8.6). These include:

- Open your online class with an introductory discussion conference in which students share their backgrounds, experiences, and personal details. The exercise—a light, nonthreatening exchange in which students become comfortable with one another—reduces the unknown and helps students learn about their teammates. It's also useful for instructors to participate in the discussion to help link students to each other.
- Solicit digital photos and post them on the class site, giving others a chance to fill out the person behind a name.
- If possible, hold a real-time conference call, welcoming students, explaining logistics and expectations, answering questions, and sharing course objectives.
- Try to create an appropriate mix of cultures in your online classes to enhance cross-cultural experiences.
- Provide a team area on the course site in which students can interact among themselves. Introduce an orientation space in the team area where members can greet one another, sharing their expectations and concerns.

TABLE 8.6. SOCIALIZATION-ORIENTATION CHECKLIST

Socialization/Orientation	Checklist
Introductory conference	☑
Posting student's pictures	☑
"Live" conference call (if possible)	☑
Multicultural team formation	☑
Create team areas	☑
Team socialization events	☑

Overlapping Communication

Owing to wide differences in expectations, English skills, learning experiences, and time orientation, it's wise to employ various media strategies to communicate key messages to students. It is often not enough to rely on e-mail messages or posted announcements alone. Faculty must introduce overlapping, repetitive messages to ensure students understand your instructions. Frequent reminders of due dates or upcoming events can be particularly helpful. It's wise to solicit feedback to guarantee that your messages have been received as you intended.

Language

Recognizing that for some, English may not be their first language, messages may be easily misunderstood, misused, or given the wrong emphasis (Payne, 2008) (see Table 8.7). With potential failures in communication in mind, the following are some suggestions:

- Keep language simple and culturally neutral. Avoid idioms, slang, colloquialisms, irony, jargon, or acronyms.
- When answering unclear questions from students, it's good to paraphrase the question to ensure you and the other students understand it.

Table 8.7. Language Checklist

Language	Checklist
Simple and clear language	☑
Use of paraphrasing for clarity of understanding	☑
"English" is not the same everywhere	☑
Use visualizations and graphics where possible	☑
Appropriate formatting (spacing, etc.)	☑
Inclusiveness in all communication	☑
Be mindful of time-zone differences	☑

- Be aware that students, even from English-speaking countries, such as Britain, Australia, and the United States, may not share the same communication patterns, meaning, or a common understanding of frequently used words or phrases.
- Whenever possible, introduce visualizations and graphics.
- Keep communications clear, simple, and unambiguous. Try to cover one point at a time, rather than mixing several ideas together.
- Use effective formatting to aid students to absorb meaning. Use of spacing, bold fonts, underlining, bullets, and other elements can help.
- Practice inclusiveness in your communication.
- Be mindful of time-zone differences and be respectful of the differences in your own expectations.

FEEDBACK

There are significant differences in how students from different cultures seek and receive performance feedback (Milliman et al., 2002). Performance (graded) feedback is a critical instructor–student exchange that can do much to foster team cohesion. The following are some best practices (see Table 8.8):

- Be timely and thorough. Students from high- and low-context cultures, for different reasons, are equally anxious to receive your feedback, especially team feedback. Returning timely response to student work—at best, within 3 days—helps students perform well.
- Your feedback should be positive in tone and constructive in nature. The following is a brief example:

TABLE 8.8. FEEDBACK CHECKLIST

Feedback	Checklist
Timely and thorough	☑
Positive in tone and constructive in nature	☑
Consistent with class and team expectations	☑
Offer prescriptive suggestions and recommendations	☑
Team feedback should be team oriented	☑
Avoid value judgments, be positive in tone, neutral in language, connected to course objectives	☑
Clear, unambiguous, and consistent	☑

Thanks, Team A, I appreciate your effort on this assignment. It is clear from your paper that you are coming together as a team. This is important because effective, positive teaming is a critical workplace competency. Being an effective teammate can help foster positive relationships and enhance your work experiences.

Your team paper was well-written, covered expected elements well, and performed the job overall. I really liked your section on X and your use of sources to add depth to your analysis.

Your good effort would have been even better with:

1. More discussion of the implications of X. What you included is fine, but it needs a bit more to cover the topic fully.
2. Add an example to clarify point X.
3. Use a bit more editing. While it was generally well done, sections X and Y are a bit disjointed. They need a stronger connection to your overall point A.
4. Introduce a stronger conclusion, recapping your key points.

Overall this is a worthy effort and I appreciate it. I am looking forward to reading your other assignments and seeing you come together even more as a team. Let me know if you need anything.

Thanks, Dr. Dool

- Your feedback should be consistent with class and team expectations.
- Whenever possible, your feedback should offer prescriptive suggestions and recommendations.

- Your team feedback should be team oriented, without identifying specific team members.

Globalization, diversity, and technology drive vast new enrollments of students into the global classroom. Many organizations now engage productive teams, particularly virtual teams, across borders. MVTs now offer faculty a unique opportunity to enhance student teaming skills to improve their eventual workplace value. MVTs bring unique challenges and opportunities and require you to be specific, consistent, and culturally aware online instructor.

ADDITIONAL RESOURCES

Online instructors may find it useful to create a common understanding of cultural bias to benchmark cultural attitudes and to underscore the need to embrace difference in multicultural teams. The following are two useful resources that may help:

Are you ethnocentric? A questionnaire designed to assess a student stereotypes about their own and other cultures (see Appendix).

Cultural intelligence. Cultural intelligence is defined as an outsider's seemingly natural ability to interpret someone else's unfamiliar and ambiguous gestures precisely the way a person in that culture might understand them (Earley and Mosakowski, 2004). This chapter offers students the opportunity to assess their cultural intelligence (see *Harvard Business Review* [October 2004], available as Reprint R0410 [www.hbr.org]).

The above resources can be used as an ice-breaking discussion in student multicultural teams.

REFERENCES

Adler, N. (2008) *International Dimensions of Organizational Behavior*, 5th ed. South-Western: Thomson Learning, Canada.

Axtell, C., Wall, T., Stride, C., and Pepper, C. (2002) Familiarity breeds content: the impact of exposure to change on employee openness and well being. *The Journal of Occupational & Organizational Psychology, 75(2)* 217.

Beckhard, R. and Pritchard, W. (1992) *Changing the Essence*. San Francisco: Jossey-Bass.

Beinhocker, E. (1999) Robust adaptive strategies. *Sloan Management Review*, 40(3), 95–107.

Chow, P. and Marcus, R. (2008) International student mobility and the United States: the 2007 open doors survey. *International Higher Education*, 50. Retrieved September 19, 2008, from http://www.bc.edu/bc_org/avp/soe/cihe/newsletter/Number50/p13_Chow_Marcus.htm.

Dadfar, H. (2001) *Intercultural Communication Theory and Practice*. Institute of Technology, Linkoping University.

Earley, P. and Mosakowski, E. (2004) Cultural intelligence. *Harvard Business Review*, October 2004.

Ford, J. (2001) *Cross Cultural Conflict Resolution in Teams*. Retrieved September 18, 2008, from http://www.mediate.com/articles/ford5.cfm.

Grzeda, M., Haq, R., and LeBrasseur, R. (2008) Team Building in an Online Organizational Behavior Course. *Journal of Education for Business*, 83(5), p. 275.

Gudykunst, W.B. (1991) *Bridging Differences: Effective Intergroup Communication*. Newbury Park, Sage Publications. Thousand Oaks, CA, p. 75.

Hamel, G. (1998) The challenge today: changing the rules of the game. *Business Strategy Review*, 9(2), 19–27.

Hira, A. (2003) *The Brave New World of International Education*. Oxford: Blackwell Publishing.

Johnson, P., Heimann, V., and O'Neill, K. (2001) The wonderland of virtual teams. *Journal of Workplace Learning*, 13(1), 24.

Lombardi, D. (1996) *Thriving in an Age of Change: Practical Strategies for Health Care Leaders*. Chicago: American College of Healthcare Executives.

Milliman, J., Taylor, S., and Czaplewski, A. (2002) Cross cultural performance feedback in multinational enterprises: opportunity for organizational learning. *Human Resource Planning*, September 2002.

Niehoff, B., Turnley, W., Hsiu, Y., and Chwen, S. (2001) Exploring cultural differences in classroom expectations of students from the United States and Taiwan. *Journal of Education for Business*, 76(5), 289.

Payne, N. (2008) Culture and the Global Team. *Ezine*. Retrieved September 18, from 2008, http://ezinarticles.com/?Culture-and-the-Global-Team&id=1311895.

Shin, T. (2005) Conflict resolution in virtual teams. *Organizational Dynamics*, 34(5), 331–345.

Sikora, P., Beaty, E., and Forward, J. (2004) Updating theory on organizational stress: the asynchronous multiple overlapping change (AMOC) model of workplace stress. *Human Resource Development Review*, 3(1), 3–35.

Sloan Consortium, (2008) *Online Nation: Five Years of Growth in Online Learning*. Retrieved September 17, 2008, from http://www.sloan-c.org/publications/survey/online_nation.

Taras, V. and Rowney, J. (n.d.) *Cross-cultural Group Management: A Review of the Research Development in the Field*. Haskayne School of Business, University of Calgary. Retrieved September 17, 2008, from http://www.ucalgary.ca/~taras/_private/Intl_Teams_Review.

United States Census Bureau (2009). US Population Projections, National Population Projections (based on Census 2000). Retrieved from: http://www.census.gov/population/www/projections/2009hnmsSumTabs.html.

Voelpel, S. (2003) *The Mobile Company, an Advanced Organizational Model for Mobilizing Knowledge Innovation and Value Creation*. St. Gallen: IFPM.

APPENDIX

ASSESSING YOURSELF

(Seton Hall University, MA-Strategic Communication and Leadership Program)

ARE YOU ETHNOCENTRIC?

Instructions All of us hold stereotypes about one group or another. This questionnaire (adapted from Gudykunst, 1991, p. 75) is designed to assess some of your stereotypes about your own and others' membership in particular co-cultures. Complete the following five steps in order:

1. Think of one co-culture of which you are a member (e.g., female, Muslim, student, or Latina)—but only select one.
2. Think of another co-culture to which you don't belong (e. g., male, Jewish, professor, or Middle Eastern American)—but again, only select one.
3. In the column labeled "My Co-culture," check five descriptive adjectives you think apply to your group.
4. In the column labeled "Another Co-culture," check five descriptive adjectives you think apply to that group.
5. Go back through the list of Descriptive Adjectives and rate each adjective you selected in terms of how favorable a quality you think it is: (5) very favorable, (4) moderately favorable, (3) neither favorable or unfavorable, (2) moderately unfavorable, or (1) very unfavorable. Put these ratings in the column labeled "Favorableness Ratings."

My Co-culture	Another Co-culture	Descriptive Adjectives	Favorableness Rating
____	____	Intelligent	____
____	____	Materialistic	____
____	____	Ambitious	____
____	____	Industrious	____
____	____	Deceitful	____
____	____	Conservative	____
____	____	Practical	____
____	____	Shrewd	____
____	____	Arrogant	____
____	____	Aggressive	____

My Co-culture	Another Co-culture	Descriptive Adjectives	Favorableness Rating
___	___	Sophisticated	___
___	___	Conceited	___
___	___	Neat	___
___	___	Alert	___
___	___	Impulsive	___
___	___	Stubborn	___
	___	Conventional	___
___	___	Progressive	___
___	___	Sly	___
___	___	Tradition loving	___
___	___	Pleasure loving	___
___	___	**TOTALS**	

Calculating Your Score The adjectives you checked reflect the stereotypes you hold about your own and another co-cultural group. Add the numbers for the two groups separately and enter the two scores below. Each score should range from a low of 5 to a high of 25. The higher the score, the more favorable the stereotype.

My Co-culture = ___ Another Co-culture = ___

Compare your two scores and consider what they say about your own degree of ethnocentrism. In what ways are you ethnocentric? To what extent do you think your stereotypes about another person's culture are real or grounded in truth? Do you think your stereotypes about your own co-culture generally reflect the way everyone really is who belongs to that group? Why or why not? Are stereotypes ever favorable? Why do you think so? Can so-called favorable stereotypes ever be a problem for members in that group? Why?

CHAPTER

9

GLOBAL CORPORATE VIRTUAL TEAMS

CHRISTINE UBER GROSSE

Standard Bank, a South African financial institution—the largest bank in Africa—operates in 18 African and 20 other countries abroad.[1] Recently, the bank launched a Global Leadership Centre in Johannesburg. As part of that effort, it asked SeaHarp Learning Solutions, a corporate training firm, to deliver a conventional classroom course, "Leading Intercultural Virtual Teams," a two-day seminar to 21 senior and mid-level managers in Johannesburg. At first, the course was delivered face-to-face, but subsequently, it migrated into an online course for virtual teams.

In addition to upholding the highest levels of integrity, serving its customers, and returning value to its shareholders, the bank identifies "working in teams" as one of its core values, noting that teams can achieve much greater objectives than

[1]Standard Bank recently announced a partnership with Industrial and Commercial Bank of China Limited (ICBC), in which ICBC became a 20% shareholder in the Standard Bank Group. ICBC is the world's largest bank by market capitalization.

Virtual Teamwork: Mastering the Art and Practice of Online Learning and Corporate Collaboration. Edited by Robert Ubell

individuals can. "We value teams within and across business units, divisions and countries," the company notes proudly.

At first, the course was offered to managers from several different units, with executives from sales, global markets, communications, finance, and global credit, drawn from South Africa, Kenya, and Swazi. Participants managed at least one intercultural virtual team, ranging from half a dozen to more than 120 members.

In the initial face-to-face class, team members reported that highlights consisted of preparing a wish list for information technology, sharing African team experiences, relating stories, problem solving, appreciating language and culture diversity, networking, and in-depth discussion.

In teams of three or four, classmates composed wish lists that imagined remote access to a home-country network, access to social networks, greater videoconferencing services, an internal Facebook, faster solutions to technology problems elsewhere in Africa, as well as access to global e-mail.

While the topic naturally lends itself to online learning, ironically, the seminar was held in a conventional, on-ground classroom for two days, a schedule that seemed overly long for extremely busy managers. Online, participants can take courses anytime, anywhere, with the freedom to work on material whenever convenient. Compared with the broad possibilities open to employees in online discussion threads, the face-to-face linear format—where only one person can participate at a time in class discussion—also seemed limiting. What's more, face-to-face learning did not lend itself to virtual team simulation.

From Face-to-face to Online

In migrating from face-to-face to online delivery, the first step was to review training materials used in the initial two-day seminar. Step two involved adapting materials for

TABLE 9.1. LEADING INTERCULTURAL VIRTUAL TEAMS

Online course units
1. Intercultural virtual teams
2. Team building
3. Meetings
4. Trust and business relationships
5. The power of diversity
6. Cultural and linguistic barriers
7. Technology
8. Team dynamics and conflict
9. Leadership development plan

asynchronous delivery, transforming in-class activities for online learning. Online modules emerged following advice from the director of Duke University's Continuing Education division to add practical tips, a unit on running virtual team meetings, and introduce situations and real problems to engage participants. (See Table 9.1 for a list of units in the online course.)

ABOUT THE COURSE

Designed for managers who work with intercultural virtual teams in business, government, and education, the online course recognizes the complexities of working in teams. It also guides managers on ways to improve virtual team effectiveness and increase productivity. Exploiting interactive learning online, participants gain insights into best practices for leading intercultural virtual teams. Topics covered include team building, developing trust, cultural and linguistic barriers to communication, team dynamics, technology, and conflict resolution.

Participants explore the pros and cons of different media for intercultural team communication, such as e-mail, telephone, videoconferencing, and face-to-face interaction. They learn practical strategies and techniques for managing intercultural virtual teams and how to apply them immediately at work. Learners prepare a Leadership Action Plan to guide continued development as leaders of virtual intercultural teams.

Self-paced and facilitator-led, the course uses a wide range of activities—blogging, podcasts, streaming videos, social-networking sites, wikis, collaborative software, and alternate reality worlds. It encourages learners to tap existing knowledge, reflect on personal experience, exchange tips, and gain new insights into leading intercultural virtual teams. Along the way, they explore new techniques and strategies for practical application on the job. (See Table 9.2 for an example of how participants reflect on their personal background and knowledge of virtual teams).

The first unit contains the following sections:

- blog/learning log;
- about you;

TABLE 9.2. BACKGROUND INFORMATION FOR ONLINE COURSE

Your virtual team

Discuss your background working in intercultural virtual teams (IVTs)

Cover the following points in a written document, audio file, or video file that you post to the course web site:

- Why are you taking the course?
- What would you like to get out of it?
- How long have you been working in intercultural virtual teams?
- How many have you worked on?
- Which team was the most successful? Why?
- Which team had the most difficulty? What happened?
- What aspect of working in intercultural virtual teams do you find most interesting?
- What is most challenging?

Learn about your teammates

Review and respond to the postings of three participants with a short comment or question

- get to know your teammates;
- your virtual team;
- challenges;
- support;
- reading and study questions.

(See Table 9.3 for sample Internet-based activities from unit 1.)

Table 9.3. Sample Online Course Internet-based Activities

Blog/learning log

Keep a blog as a learning log or journal to record your thoughts, insights, and information. Keep track of how you apply new ideas to leading your team, and its impact on team outcomes. Post your blog on the course web site, making new entries each time you work on the course. Follow blogs of other participants to see how they think and how they apply new tips and strategies

About you

One of the keys to leading an effective virtual team is getting to know your teammates well. Introduce yourself to the team in one of the following three ways:

1. Write about yourself on the course web site
2. Create an audio file and upload it to the course site
3. Shoot a video and upload it to the site

Discuss some of the following points in your introduction:

- Your name, title, and what you do at your company
- Your hobbies and interests outside of work
- Your educational background
- Your family
- Your career history

Know your teammates

Review your teammates' introductions and respond to three of them with a brief comment or question

GLOBAL COMMERCE

Global companies increasingly rely on intercultural virtual teams to complete short- and long-term projects. These teams present unique challenges for managers, especially building trust, communicating across language and cultural barriers, and using technology effectively.

"Virtual" refers to the electronic communication of team members. Technology-enabled communication allows teams to function, even when physically separated by distance and time zones. Global trade and technology have changed the work environment for organizations of all sizes. With technological developments in communications and logistics, smaller companies can now compete in the international marketplace.

Global teams often face complex challenges that do not allow them to function as well as expected. This course addresses how managers deal with the challenges in leading effective teams. (See Table 9.4 for an example of how learners reflect on personal challenges in the online module).

Virtual team members need to choose appropriate communication channels for their purposes and have to balance distance work with face-to-face communication. Effective team

TABLE 9.4. CHALLENGES FACED BY VIRTUAL TEAM MEMBERS

What challenges have you faced in leading your virtual team? Describe challenges you have come across in the following areas, if applicable:

1. Communication
2. Trust
3. Technology
4. Language and cultural differences
5. Leadership
6. Team dynamics
7. Conflict resolution

Review teammate challenges: Consider challenges faced by at least two other participants in the course. Compare these with the challenges you have faced. Write a short comment or question in response to the two teammates.

leaders encourage open communication and brainstorming, and avoid assignment of blame. They know how to develop strong relationships built on trust and understanding. Team leaders actively show respect for other cultures and interest in other languages. In their communication and actions, they demonstrate knowledge of how diversity strengthens their team.

Broadly speaking, team building has four components—forming, norming, storming, and performing. In virtual teams, team formation plays a critical role in overall team effectiveness. The difficulty of working at a distance presents a variety of challenges that involve time, distance, technology, isolation, trust, and cultural and linguistic barriers to communication. It is essential for team members to get to know one another as rapidly as possible.

Nokia encourages virtual teams with several members who already know one another from previous work engagements. Previous knowledge, existing relationships, and trust found in teams with previous connections have shown to be invaluable in team outcomes. Whenever possible, it's best to capitalize on existing relationships. For those unfamiliar with the rest of the team, it is vital for them to bridge distance by developing online personal relationships as quickly as possible.

Virtual team leaders can build close personal relationships and trust by showing interest in team members, by encouraging frequent informal communication, and by providing as many opportunities for face-to-face interaction as possible. Virtual teaming improves significantly when participants are brought together at the start. Informal interaction generates richer social networking if members meet face-to-face at first. If gathering together in the same place is not feasible, it's best to introduce videoconferences, or video cameras connected to Skype. Table 9.5 offers an exercise in how to encourage manager support for their teams.

TABLE 9.5. SUPPORT FOR MEMBERS OF VIRTUAL TEAMS

Discuss the following questions with a classmate or review with another member over Skype or by telephone:

1. Where do you go for help managing your virtual intercultural team?
2. Who provides technical support?
3. Do you have a mentor or another manager you can go to for advice about managing your team? If not, how might you go about finding someone to advise you?
4. Describe your relationship with tech-support. How might you improve it?

Developing trust, as well as managing quality business relationships, is a critical skill that virtual team leaders must absorb. While mutual trust plays an essential role in building successful international alliances and teams, managers often admit that trust is not an easy attribute to achieve online. Trust is a necessary condition for successful teamwork, especially in virtual teams in which there is uncertainty and incomplete knowledge of everyone in the group (Child, 2001). Jarvenpaa et al. (1998) find that trust among global team members depends on the perception of each other's ability, integrity, and benevolence.

To encourage open communication and brainstorming, members must avoid blaming one another. It's best if participants address problems together and share responsibility, not only for problems encountered, but also for successes achieved.

TIPS FOR DEVELOPING TRUST ONLINE

- Communicate openly and frequently.
- To receive trust, give trust.
- Be honest.

- Establish strong business ethics.
- Do what you say you will do.
- Be consistent and predictable.
- Set a tone from the start that supports future interaction.
- Be accessible and responsive.
- Keep confidences.
- Create social time for team interaction.

THE POWER OF DIVERSITY

High-quality managers recognize that diversity strengthens intercultural teams. They encourage each member to contribute, acknowledging that each one brings different personal and cultural perspectives to the task, enriching and extending the work of the team.

It's good to recognize that different cultures reflect varying responses to scheduling, goal setting, and task assignment (Milosevic, 1999). Diversity brings a broader range of expertise, resources, and viewpoints to projects (Townsend et al., 1998). In business communication classes, students often appreciate how members from other cultures add different perspectives, helping them to see things from a different angle. Diversity stimulates new ideas and enhances creativity. Working in diverse teams often allows participants to appreciate different points of view and learn how other cultures behave, often erasing stereotypes.

Varner (2001) emphasizes other important cultural effects on team dynamics, such as giving and receiving criticism and feedback, willingness to participate and offer ideas, and contradicting superiors.

Getting to know everyone's strengths and background at the beginning of each project can be very productive. When you know about your teammates, your team can achieve success by playing to one another's strengths while minimizing weaknesses. When members recognize each other's expertise, experience, skill, and capabilities, teams tend to distribute work more equitably. For example, a team member who demonstrates excellent organizational skills can be enlisted to monitor project completion. Those with good time management skills can keep members on track. Others with interpersonal skills may be successful in helping resolve conflict and build consensus following disagreement.

Graduate business students at Thunderbird's School of Global Management identified distinct advantages in working in multicultural teams (Grosse, 2002). Most described the experience as extremely positive. They reported that it was fun as well as refreshing to learn about other cultures while getting the job done. As a byproduct, they learned how to communicate more effectively with people from all over the world. They exchanged information beyond the scope of the project, and broadened their knowledge of different cultures and places. As an added benefit, they established new friendships. Enjoyment of working in a team was one of the three key factors that Snow et al. (1996) identified as critical for success in a transnational team. Two other key factors are commitment to the team's mission and dependability in performing assigned tasks.

LANGUAGE AND CULTURAL BARRIER TIPS

- *Communicate continuously.*
- *Engage in active listening.*
- *Keep communication simple and clear.*
- *Exploit appropriate technologies.*
- *Build relationships and trust.*

- *Show respect for other cultures.*
- *Be sensitive to cultural differences.*
- *Ask for clarification.*

COMMUNICATING ACROSS CULTURES

To communicate effectively across cultures, managers must practice intercultural sensitivity, good relationships, and trust. Ackley and Barker (2001) identify four criteria to assess intercultural sensitivity: (1) positive feelings toward interactions with people from other cultures; (2) positive feelings of those from other cultures toward the individual; (3) successful job completion; and (4) freedom from culture-contact stress.

Some find it challenging to work with people whose native language is different from their own, remarking that teammates often misunderstand what is said, sometimes concluding something entirely different than what is intended. Some US students worry that those from other cultures may be offended by American openness and directness, an observation that supports Thomas's (1999) belief that team members need to understand how cultural differences affect team dynamics.

Thunderbird's business communication teams identified three basic ways to overcome communication obstacles—patience, respect, and attentive listening (Grosse, 2002). Participants found that being patient in dealing with difference and being open-minded and respectful facilitated communicating across cultures. It took time and patience to recognize and adapt to different styles, but listening to each other patiently and attentively helped overcome many communication problems. Optimally, it's best to let your teammates know that you understand and appreciate their points of view. But when you fail to understand, it's prudent to persist in asking for clarification to avoid continued misunderstandings.

Common goals—such as performing well on your team assignment—also contribute to forming a united team. Others found it helpful to demonstrate a willingness to work things out and keep active participation going. In the end, most agreed that they learned how to adopt strategies designed to communicate effectively in a challenging environment. They succeeded in improving their listening skills as well as executing better methods for sharing ideas.

While most acknowledged that working in a team can take more time and effort to accomplish their goals than doing things on their own, nonetheless, they conceded that the benefits outweighed the disadvantages. Most agreed that the finished product was of greater quality and had less chance of error than if they had performed the work alone.

Being open-minded about other cultures goes a long way to establishing rapport. Awareness of differences in cultural values and beliefs, communication styles, approaches to decision-making, problem solving, and conflict resolution help teams overcome cultural obstacles. Team members can help break down language barriers by showing respect for other cultures and languages. To enhance team building, John Purnell, formerly at Digital Corporation, recommends that global participants show appreciation for cultural differences, acknowledge the value of colleague's time, hone their listening skills, learn how to resolve conflict, know how to plan projects, and enhance their computer expertise (Odenwald, 1996). Iles and Hayers (1997) believe that transnational project teams must also learn how to negotiate and how to think strategically.

Obstacles to effective communication often occur at the interpersonal level, with lack of appreciation for what kinds of behaviors are acceptable and what kinds of interactions may be offensive. Some have difficulty appreciating what level of directness is appropriate. Others may be too informal with people from cultures who prefer more formal address. Still others may be unaware that certain cultures cannot tolerate especially aggressive or argumentative confrontation.

Effective managers have found that continuous communication is the key to success. They often exchange daily e-mails with their team members. At UPS, for example, the director of international public relations holds a weekly voice conference with his regional directors around the world.

Communicating across Languages and Cultures

- *Build a network of relationships.*
- *Understand how diversity strengthens a team:*
    - *Explore the pros and cons of intercultural teams.*
- *Build trust and understanding.*
- *Be open to learning about other cultures:*
    - *Acknowledge different cultural values, beliefs, and communication styles.*
    - *Appreciate different approaches to decision-making, problem solving, and conflict resolution.*
- *Balance distance activities with face-to-face time.*
- *Show respect for other cultures and languages:*
    - *Overcome cultural differences.*
    - *Break down language barriers.*
- *Use appropriate communication channels.*

Using Technology in Cross-cultural Teams

While all virtual teams depend on technological intervention, cross-cultural teams face special challenges.

E-MAIL

E-mail provides certain advantages for virtual teams composed of members with different native languages and cultures. It allows time for people to compose and process their messages, reducing the pressure to communicate immediately as one does in person or by telephone. With e-mail, participants have time to edit what they plan to write before sending a message. The receiver also has time to process it and respond appropriately. Non-native speakers who join virtual teams often feel they can communicate more effectively with e-mail than with other forms of communication because it allows them to compose and respond with more consideration and reflection.

E-mail also makes it easier to ask for clarification without losing face. The apparent spontaneity and immediacy of e-mail helps teams communicate more fluently and since e-mail messages are typically short, non-native speakers find it easier to comprehend and post.

TELEPHONE

The phone is a difficult way to communicate, because visual cues are missing. It is especially hard to talk with people from different cultures or nationalities who are non-native speakers.

VIDEOCONFERENCE

For a large group, video has advantages. On the phone, it is difficult to know who is saying what, especially when more than two or three people are speaking. For a large-group discussion, videoconferencing works better.

FACE-TO-FACE

Face-to-face meetings bring a special chemistry and power to communication. Executives typically acknowledge the advantages of face-to-face. But for many virtual teams, face-to-face encounters are not possible.

By promoting the advantages of virtual intercultural teams, team leaders can overcome many challenges posed by cross-cultural communication and technology. They can take advantage of the opportunities that derive from working multiculturally. With open minds and respect for other languages and cultures, managers can help teams avoid misunderstandings. Showing patience, care, and sensitivity, managers can lead their teams across language and cultural barriers to more effective communication.

Appreciating how to communicate effectively helps managers to achieve higher performance and avoid costly delays. By their own example, managers can enhance team members' appreciation for cultural diversity. As managers exploit the power of technology to communicate with their teams across time and space, they build strong human relationships that may never have occurred otherwise.

REFERENCES

Ackley, J. and Barker, R. T. (2001) The presence of five intercultural dimensions in Mid-Atlantic managers: an exploratory study. *Paper presented at the 2001 ABC European Convention*, Dresden, Germany, May 24–26.

Baker, S. and Heather, G. (2008) Beyond blogs. *Businessweek*, June 2.

Barrett, D. J. (2001) Teaching MBA students to manage intercultural teams. *Paper presented at the 2001 ABC European Convention*, Dresden, Germany, May 24–26.

Child, J. (2001) Trust—the fundamental bond in global collaboration. *Organizational Dynamics*, 29(4), 274–289.

Conlin, M. (2008) The waning days of the road warrior. *Businessweek*, June 2.

Covey, S. M. R. (2006) *The Speed of Trust*. New York: Free Press.

Daly, C. B. (1996) Does diversity matter? *Harvard Business Review*, 74(3), 10–11.

DiStefano, J. J. and Maznevski, M. L. (2000) Creating value with diverse teams in global management. *Organizational Dynamics*, 29(1), 45–64.

Duarte, D. L. and Snyder, N. T. (2006) *Mastering Virtual Teams, 3rd ed*. San Francisco, CA: Jossey-Bass.

Gratton, L. and Erickson, T. J. (2007) 8 ways to build collaborative teams. *Harvard Business Review*, (November), 100–109.

Grosse, C. U. (2002) Managing communication within virtual intercultural teams. *Business Communication Quarterly*, 65, 22–38.

Hamm, S. (2008) International isn't just IBM's first name. *Businessweek*, January 28, pp. 34–40.

Iles, P. and Hayers, P. K. (1997) Managing diversity in transnational project teams. *Journal of Managerial Psychology*, 12(1/2), 95–118.

Jarvenpaa, S. L., Knoll, K., and Leidner, D. E. (1998) Is anybody out there? Antecedents of trust in global virtual teams. *Journal of Management Information Systems*, 14(4), 29–64.

Keenan, F. and Ante, S. E. (2002) The new teamwork. *Businessweek*, February 18, EB 12–16.

Kostner, J. (2001) *Bionic eTeamwork*. Chicago, IL: Dearborn Trade Publishing.

Lowry, T. and Balfour, F. (2008) It's all about the face-to-face. *Businessweek*, January 28, pp. 48–51.

McCain, B. (1996) Multicultural team learning: an approach towards communication competency. *Management Decision*, 34(6), 65–69.

Milosevic, D. Z. (1999) Echoes of the silent language of project management. *Project Management Journal*, 30(1), 27–40.

Odenwald, S. (1996) Global work teams. *Training and Development*, 50(2), 54–57.

Overholt, A. (2002) Virtually there? *Fast Company*, 56, 109–115.

Snow, C. C., Snell, S. A., and Davison, S. C. (1996) Use transnational teams to globalize your company. *Organizational Dynamics*, 24(4), 50–67.

Thomas, D. C. (1999) Cultural diversity and work group effectiveness. *Journal of Cross-cultural Psychology*, 30(2), 242–264.

Townsend, A. M., DeMarie, S. M., and Hendrickson, A. R. (1998) Virtual teams: technology and the workplace of the future. *The Academy of Management Executive*, 12(3), 17–29.

Varner, I. (2001) Successful intercultural work teams: the role of human resource management. *Paper presented at the 2001 ABC European Convention*, Dresden, Germany, May 24–26.

Warkentin, M. E., Sayeed, L., and Hightower, R. (1997) Virtual teams versus face-to-face teams: an exploratory study of a web-based conference system. *Decision Sciences*, 28(4), 975–996.

Yates, J. and Orlikowski, W. (2002) Genre systems: structuring interaction through communication norms. *The Journal of Business Communication*, 39(1), 13–36.

CHAPTER

10

CORPORATE VIRTUAL TEAMING

LUTHER TAI

Consolidated Edison Company of New York, Inc., in a partnership with Stevens Institute of Technology's online learning unit, WebCampus, offers an online course, "Analytical Capabilities for Business Improvement." At the start in 2005, Stevens interviewed Con Edison senior managers on company needs and proposed an online training solution.

Con Edison's aim was to have its employees become more analytical in problem solving. The goal was to use e-learning to engage staff in different company settings to work in teams on real problems facing the company. Stevens, together with Con Edison, customized an e-learning effort to provide Con Edison employees with tools and processes needed to conduct analyses for structured decision-making for business improvement. Employees participate in virtual classrooms and engage in team discussions and problem-solving techniques from any location: work, home, or even traveling. Users can conveniently access e-learning sites from anywhere, at any time.

In their course, Con Edison employees attack actual business cases with analytical tools that could be applied directly to

Virtual Teamwork: Mastering the Art and Practice of Online Learning and Corporate Collaboration. Edited by Robert Ubell

problems that they struggle with every day. For many, virtual teaming itself is an entirely new learning experience. For presentation, posting, and communication with their instructors and with each other, Stevens facilitates asynchronous collaboration from a web portal. Participants also attend webcasting sessions in real time from distant locations. The approach is especially useful for employees who work at different sites and on different schedules.

The true benefit of working virtually is the flexibility it gives employees to complete projects and assignments from anywhere at anytime. It allows participants to maintain their own schedules, allowing them to study at their own convenience and to communicate when opportune. Online interaction permits teams to function as if employees are all together in the same space. To assess conclusions and ideas generated, other teams provide feedback to achieve a solid check and balance. What's more—and perhaps equally of value—intense online interaction builds strong relationships that often continue beyond the virtual classroom.

Moving away from a physical to a virtual classroom requires a great deal of adaptation. At first, participants who have never experienced virtual reality are unnerved by online protocols. Over time, however, they become familiar and comfortable with e-learning and use the Internet as a link between instructors and teams and among team members themselves. To ensure deliverables meet due dates and course schedules, employees generally find that participation demands a greater degree of self-discipline than they had expected.

While 17 teams have so far collaborated on projects using the define–measure–analyze–improve–control (DMAIC) process, this chapter describes a few examples from a wide range of company problems including Help Desk, recruitment, violations, construction, and gas mapping. The experiences and processes encountered by these teams have been indicative of all teams.[1]

HELP DESK CALL VOLUME REDUCTION

Con Edison's Help Desk provides first-level support to employees and contractors for all information technology problems at the company, including desktop and peripherals, servers and voice and data telecommunication services, and application support. The Help Desk is staffed seven days a week, 24 hours a day.

During a five-year period, call volume significantly increased, from 69,000 calls in 2001 to 83,000 in 2005. The length of each call also increased; the average call in 2005 lasted more than 6 minutes. Opportunities to provide more value-added services to employees and contractors needed to be explored, but the methods to deliver them had to be achievable.

A team comprised of employees from Information Resources, Facilities, Electric, and Central Services tackled this issue. The project team used a process map (Figure 10.1) to outline the current process and analyze system capabilities.

The team analyzed the procedure used when an employee places a call:

1. The caller places a call to the Help Desk.
2. The Automatic Call Distributor (ACD) answers the calls, provides advisory information, and presents the

[1]Other teams analyzed steam resource planning, New York independent system operator billing process, self-service applications, dielectric fluid loss, training course management, system reliability improvements for Manhattan bank transformer process, customer service phone calls, billable demand meters, restoration process of secondary mains on the electric distribution system, leadership development needs analysis, reduction of sulfur hexafluoride (SF_6) release into the environment, and payroll change authorizations.

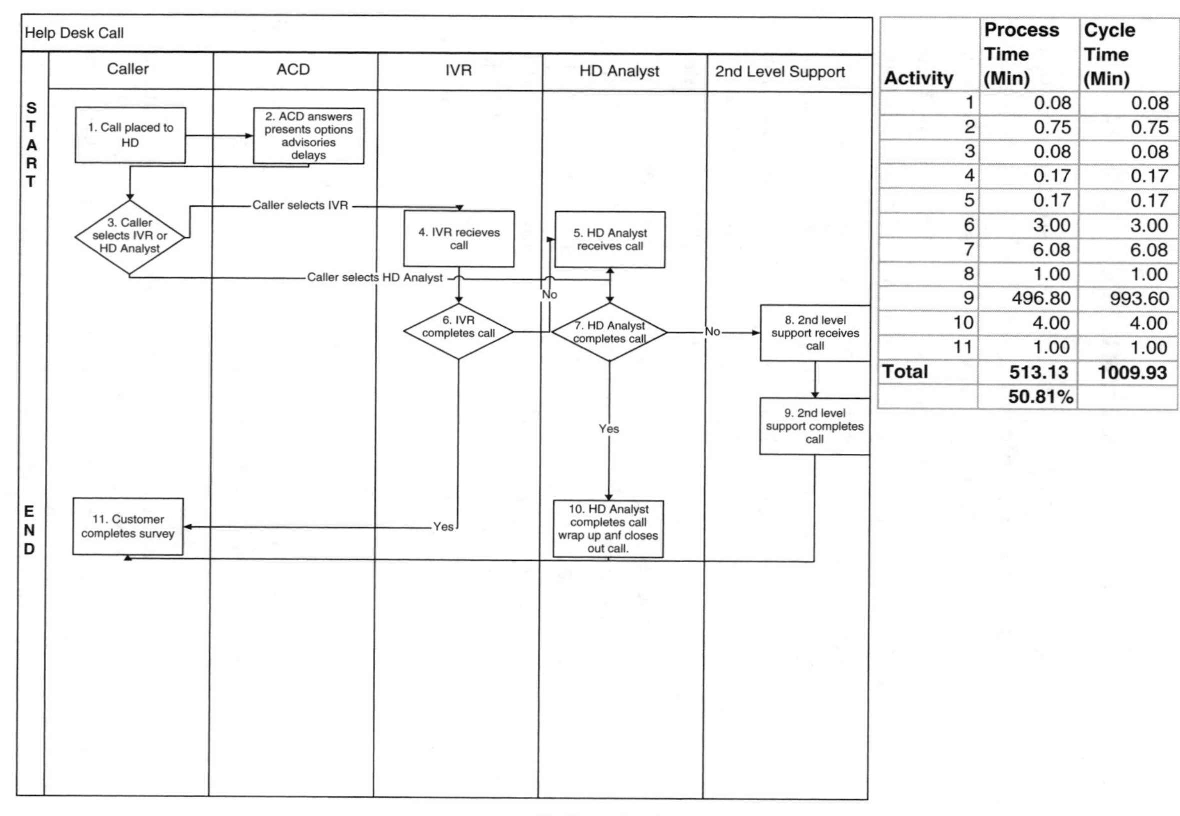

Activity	Process Time (Min)	Cycle Time (Min)
1	0.08	0.08
2	0.75	0.75
3	0.08	0.08
4	0.17	0.17
5	0.17	0.17
6	3.00	3.00
7	6.08	6.08
8	1.00	1.00
9	496.80	993.60
10	4.00	4.00
11	1.00	1.00
Total	**513.13**	**1009.93**
	50.81%	

FIGURE 10.1. Call analysis process map.

option of using the Interactive Voice Response (IVR) unit or speaking with an analyst.

3. The caller decides whether to have IVR process the call or speak with an analyst.
4. If the caller chooses the IVR option, IVR receives the call.
5. If the caller decides to speak with an analyst, an analyst receives the call.
6. If IVR resolves the call, the caller completes a satisfaction survey (step 11). If IVR does not resolve the call, it is routed to an analyst (step 5).
7. If an analyst resolves the call, the caller completes a satisfaction survey (step 11). If an analyst does not resolve the call, it is escalated to a second-level support group. The analyst informs the caller that the call has been escalated.
8. The second-level support group receives and assesses the call.
9. The second-level support group resolves the call.
10. The caller receives a satisfaction survey.
11. The caller completes the satisfaction survey.

Studying the process map, the team analyzed the number of calls handled by the IVR or calls processed by an analyst, the average length of a call, and the abandoned call rate. The team also reviewed surveys to assess how well or poorly the Help Desk processed calls. The team prepared a Pareto chart of problems found from the process map (Figure 10.2). The chart was used to analyze and identify key problems.

The team also created a cause and effect diagram to further analyze issues (Figure 10.3). The diagram identified key drivers for calls and potential areas for improvement.

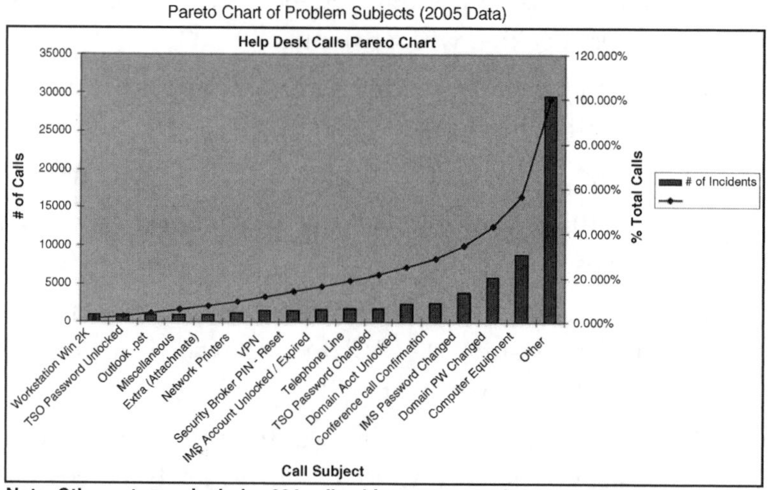

FIGURE 10.2. Pareto chart of Help Desk problems.

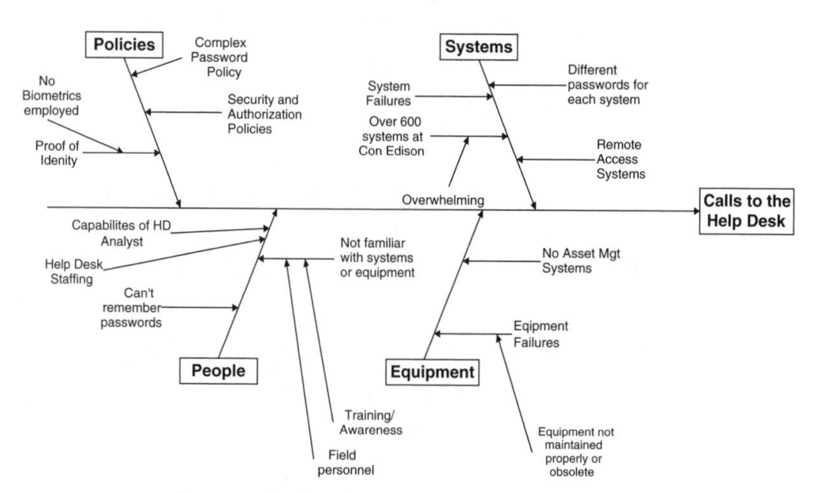

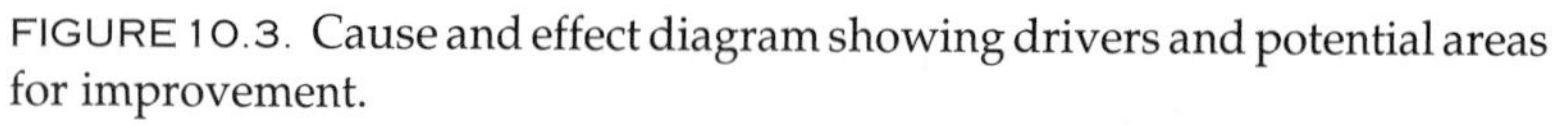
FIGURE 10.3. Cause and effect diagram showing drivers and potential areas for improvement.

The team concluded that the Help Desk should continue to provide first-level support for all information technology-related problems, but it needed to find ways to provide more value-added services. The objectives were to improve analyses of all calls, increase self-service capabilities, and reduce call volume and call time. The team proposed employing the following options:

1. Methods to reduce computer equipment calls by at least 10%:
   - trend equipment problems and failures;
   - eliminate problem or obsolete equipment;
   - train personnel on proper use of equipment;
   - provide a self-service reporting function.
2. Methods to reduce domain and other password-related calls by at least 10%:
   - provide a self-service reset function on IVR and on a self-service intranet portal, resulting in significant reduction in call cycle and process time;
   - provide training and other awareness programs.
3. Methods to reduce average call time:
   - analyze call time drivers (subjects with longest call time);
   - provide standard resolutions for common problems;
   - reduce call subject categories (currently more than 700);
   - provide self-service reference materials;
   - formally train Help Desk analysts;
   - staff for peak volumes on Monday and Tuesday.

Once the team prepared its analysis and proposed a plan to improve the Help Desk, a development phase was introduced, including designing and building a self-service intranet portal. The portal would help eliminate calls by allowing users to reset a password, process a trouble call, query the knowledge base, or access training materials (Figure 10.4).

The team recommended that the Help Desk continually perform call analyses to ensure that the self-service portal operates efficiently, to identify techniques to prevent calls, and to guarantee adequate staff for peak periods. Ongoing customer service training programs and continuing "lunch and learn" seminars would improve awareness and provide pertinent training techniques to Help Desk personnel. The team also recommended training and improved awareness of various self-service pathways be delivered to users.

The team continues to monitor the services that have been adopted. Remarkably, all major metrics were exceeded as early

IR Help Desk Self Service Portal

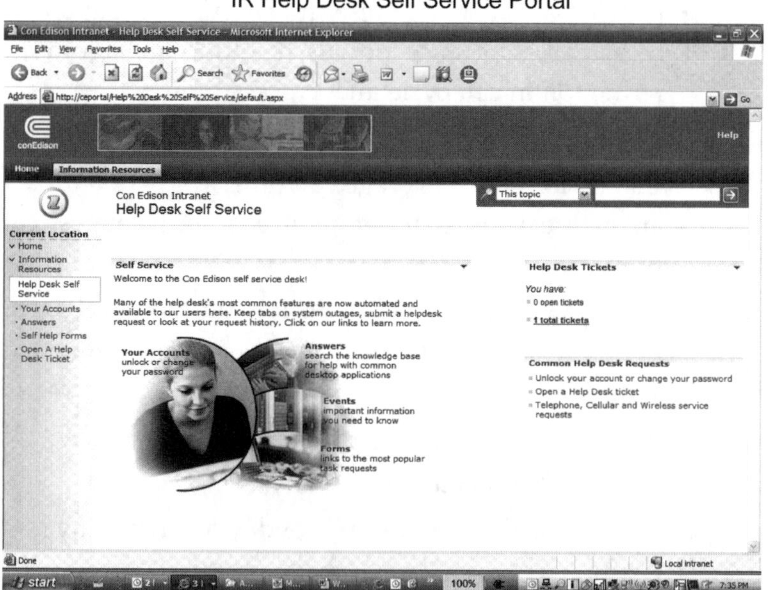

FIGURE 10.4. Help Desk self-service intranet portal.

as the first half of 2006. Less than 6% of calls were abandoned, more than 75% of calls were answered in 24 seconds or less, and more than 80% of calls were resolved by the Help Desk. As of August 2008, the abandoned call rate was merely 3.3%, 87% of calls were answered in 24 seconds or less, and 89% of calls were resolved by the Help Desk.

As more employees became accustomed to the self-service intranet portal, the number of calls per day declined. In 2006, the number of calls per day averaged 261, in 2007 it was 226, and, as of August 2008, the average number per day was 198. In 2008, there were 20,470 password-related incidents; 17,350 were processed using the portal, while only 3120 were generated from the IVR system.

RECRUITMENT PROCESS

Recruitment encompasses all steps used to attract and employ the most qualified candidates. At Con Edison, the process begins with a request from an internal organization to hire a new staff member. Once a request is made, the job is posted, applicants are pre-screened, and interviews are conducted. Soon afterwards, recommendations are made and the selection phase begins.

At each stage, participation from the hiring organization is required, but levels of responsibility fluctuate with each task. While the Human Resources department is the process owner, it partners with each hiring organization. As a company with approximately 15,000 employees and aggressive recruitment goals for the future, each hiring organization has a responsibility for different steps in the entire process.

A virtual team drawn from Con Edison's Human Resources department outlined the company's current recruitment process. The team identified improving the speed of management and union recruitment as the company's key priority while still maintaining quality and controlling costs.

During a root cause analysis, the team reviewed the current process and held several meetings with subject-matter experts to evaluate each step of the recruitment process. As a result, the team selected specific metrics to truly evaluate the process—requisition turnaround (the time it takes from requisition to candidate hire), cost per hire, and quality of hire.

Using a fishbone diagram (Figure 10.5), a cause and effect tool that points to possible causes for an effect or problem, the team identified three factors that might be contributing to long turnaround time—the hiring organization, workflow, and systems. As expected, participants found bottlenecks contributing to wide gaps between process and cycle times. Exploiting the group's expertise, the team held several brainstorming sessions to sort out ideas into useful categories.

The team identified several potential efficiencies that might improve recruitment cycle time. One efficiency option employed e-mail to communicate with potential candidates instead of intensive phone work. Interviews, rejections, and

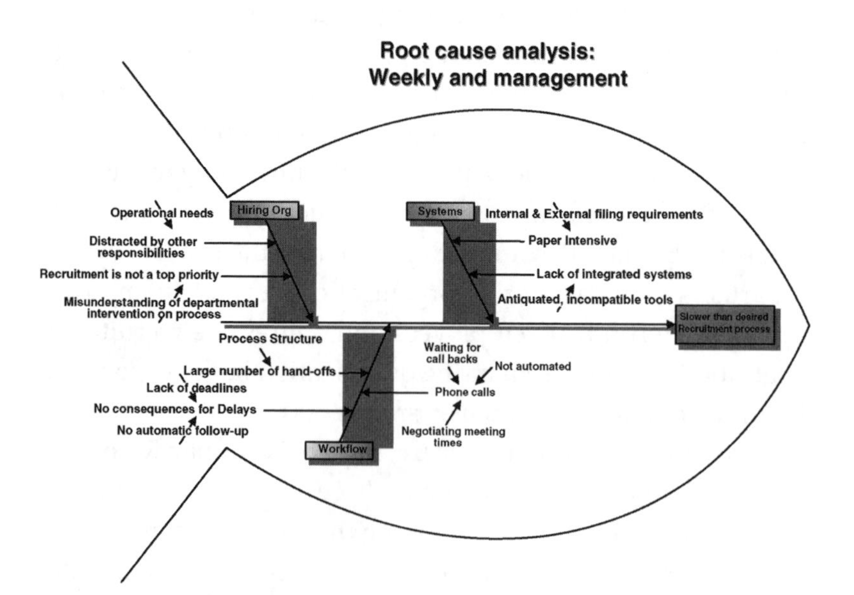

FIGURE 10.5. Fishbone diagram for root cause analysis.

position offers could be accomplished in this manner. The team also proposed that certain responsibilities be shifted to other departments, increasing stakeholder participation. Overall, the team concluded that the recommended improvements may not only increase efficiency, but also improve candidate perception of the position.

Ultimately, the team highlighted three major areas for improvement—technological enhancements, procedural revisions and standardization, and recruitment process controls. In addition to the increased use of e-mail, the team recommended designing and implementing a program tentatively called eOrientation; this program would replace a one-day classroom employee orientation session. The team also suggested enhancing the online job application form with an automated matrix-filtering mechanism to better identify qualified candidates.

The team also proposed that standard language be used for job postings and recommended better process controls, including periodic reviews of metrics, cycle, and process times at each stage. The team is currently working on these changes and expects to complete the project in 2009.

REDUCING NOTICES OF VIOLATION

Con Edison operates transmission and distribution systems that deliver electricity, natural gas, and steam to various customers in the five boroughs of New York City and in Westchester, Orange, and Rockland Counties. Within New York City, the operation and maintenance of these systems require opening and entering subsurface structures such as manholes and vaults and excavating within the city streets' right-of-way. In order to complete this work, New York City Local Laws require that permits be obtained for specific tasks; these permits include work rule stipulations that must be followed, including returning that roadway and/or subsurface structure to its original condition once work is completed. The New York City Department of Transportation (DOT) has jurisdiction over these matters. Noncompliance of work rules results in a notice of violation (NOV). Because of the complications involved in utility work on and under the streets of the city, Con Edison receives thousands of such notices annually and, consequently, pays significant penalties (Table 10.1).

Employees drawn from Con Edison's departments of Construction, Gas Operations, Electric Operations, and Public Affairs collaborated to find ways to reduce the number of violations and associated penalties. They charted the internal processing flow of any type of violation to understand how various internal departments interact and intersect with each other (see Figure 10.6). Using data collected by Central Compliance, the team reviewed the NOVs received over a three-year period, separating violation types, and total of each received and associated fines. The team calculated the relative importance of each violation type by using a House of Quality diagram (see Figure 10.7) and a Pareto chart (see Figure 10.8). After reviewing the results, the team selected a particular violation type that affected multiple operating organizations: "No Notice Prior to Starting Work on a Protected Street." These infractions occurred when the company failed to notify the DOT before backfilling an excavation on a "protected street."

TABLE 10.1. NOVS RECEIVED BY CON EDISON AND RESULTING FINES

Year	NOVs	Penalty (million $)
2000	3172	1.4
2001	6480	2.6
2002	6966	3.2
2003	8737	5.1
2004	7224	3.5
2005	5929	4.7
2006	5232	4.0

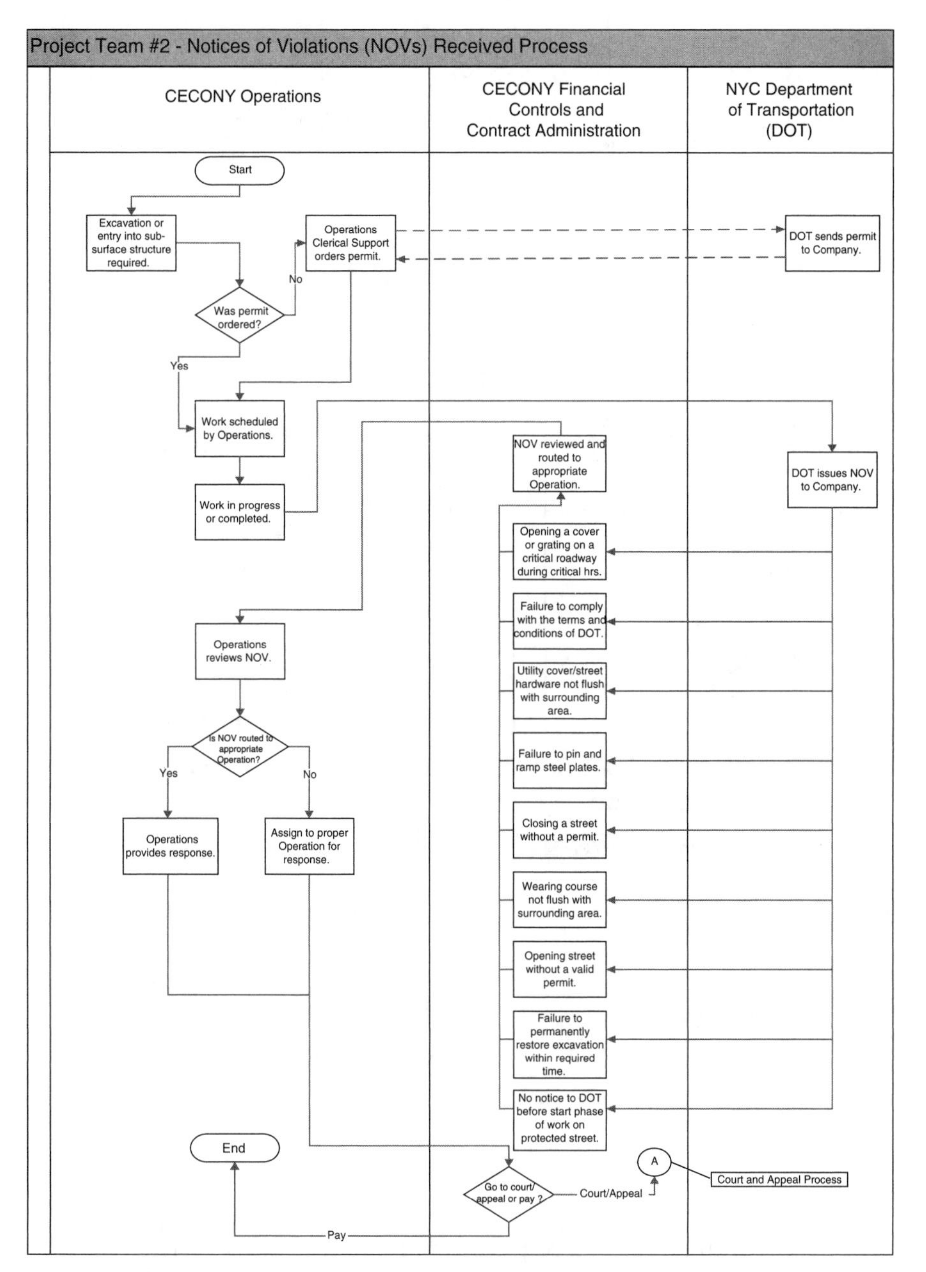

FIGURE 10.6. Internal workflow for NOV processing.

Customer = Company (Con Edison)

Roof Inter-Relationship	
Symbol	Relationship
⊙	Strong Positive
O	Positive
X	Negative

Customer Needs/Requirements (Technical Requirements →)	Emergency Permit Violations	Hardware Not Flush with Surrounding Areas	No Notice to DOT Prior to Work on a Protected St.	Working out of Stipulations	Areas Not Flush with Surrounding Roadway	Importance to Customer (1-5)	Current Performance	Desired Performance	Improvement Level	Absolute Overall Weight	Normalized Priority
Maintain Traffic Flow	1	9	1	9	9	3	3	3	1.00	3.00	0.03
Improve Quality of Roadway	1	9	9	1	9	3	1	5	5.00	15.00	0.13
Reduce Need for Additional Permits	3	3	3	3	3	5	1	3	3.00	15.00	0.13
Reduce NOV Penalties	9	9	3	9	3	5	1	5	5.00	25.00	0.21
Minimize Rework			9			5	3	5	1.67	8.33	0.07
Prevent Escalating Violation Costs	1	9	1	1	9	3	3	5	1.67	5.00	0.04
Minimize Lawsuits	9	9			3	5	3	5	1.67	8.33	0.07
Minimize Negative Press	9	9	9	9	9	5	3	5	1.67	8.33	0.07
Reduce Effect on Rate Case	3	1	1	1	1	5	3	5	1.67	8.33	0.07
Improve Vehicle Safety (Public)	9	9		3	3	5	3	5	1.67	8.33	0.07
Improve Pedestrian Safety (Public)	9	9		3	3	5	3	5	1.67	8.33	0.07
Improve Employee Safety	3					5	5	5	1.00	5.00	0.04
Total Points	5.4	6.7	3.6	3.8	4.1		23.5			118.00	1.00
% of Total Points	0.2	0.3	0.2	0.2	0.2		1				
Total Sum of All Projects	21										

Mapping Matrix Relationship	
Relationship	Weight
Strong	9
Medium	3
Weak	1
None (blank)	0

Priority of Customer Needs	
Priority	Weight
High	5
Medium	3
Low	1

FIGURE 10.7. House of Quality diagram for notices of violation.

Backfilling consists of packing and compacting an excavation. Protected streets are those that have been completely repaved within the last five years. According to the DOT rules, a city inspector must be present to ensure that backfilling on

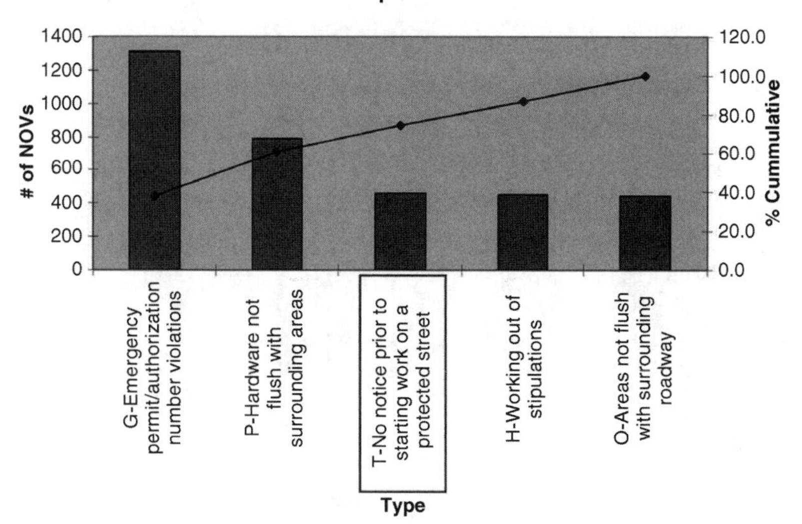

FIGURE 10.8. Top 5 DOT notices of violation over a three-year period.

protected streets is performed according to specifications. The city's transportation department requires that it be notified at least 2 hours before backfilling begins. Failing to call for an inspector can result in a NOV with a fine of $250.

Once the specific violation type was identified, the team processed a three-year dataset into a histogram to discriminate the number of these violation types by borough (see Figure 10.9). Using a second histogram, they differentiated the number of notices sent to each operating organization (see Figure 10.10).

From a macroscopic viewpoint, the team identified key elements in the current permit request process and discovered that many potential flowpaths exist. Each flowpath differed depending on who requested the permit, communication between parties, and whether Con Edison or a contractor performed the work. Gathered data were then inputted into a cause and effect diagram to isolate root causes behind the excessive

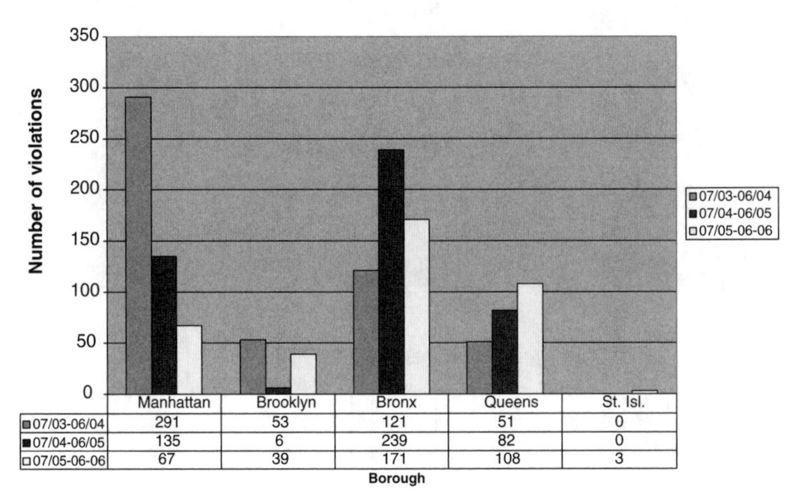

	Manhattan	Brooklyn	Bronx	Queens	St. Isl.
07/03-06/04	291	53	121	51	0
07/04-06/05	135	6	239	82	0
07/05-06-06	67	39	171	108	3

FIGURE 10.9. Notice of violation by borough.

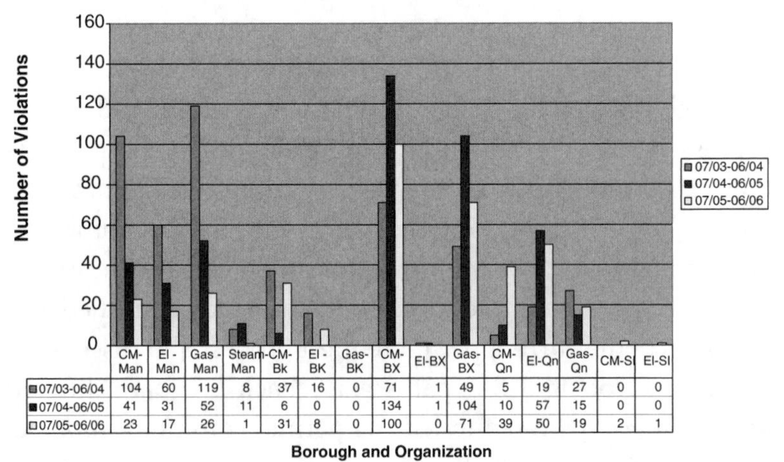

	CM-Man	El - Man	Gas - Man	Steam Man	CM-Bk	El - BK	Gas-BK	CM-BX	El-BX	Gas-BX	CM-Qn	El-Qn	Gas-Qn	CM-SI	El-SI
07/03-06/04	104	60	119	8	37	16	0	71	1	49	5	19	27	0	0
07/04-06/05	41	31	52	11	6	0	0	134	1	104	10	57	15	0	0
07/05-06/06	23	17	26	1	31	8	0	100	0	71	39	50	19	2	1

FIGURE 10.10. Received notice of violation by operating organization.

number of violations. Three key areas emerged—responsibility or accountability, communication, and process and uniformity (see Figures 10.11 and 10.12).

The team methodically analyzed each root cause and discovered a cyclical and intertwining relationship between them. Work at a job site can be accomplished in three different ways by varying crews. In one combination, a contractor initiates work (such as digging and opening a street or substructure) followed by a Con Edison crew that performs detailed labor. The Con Edison crew is then replaced by the returning contractor who backfills and repaves as necessary, returning the site to its pre-work condition. Second, the work is performed entirely by a

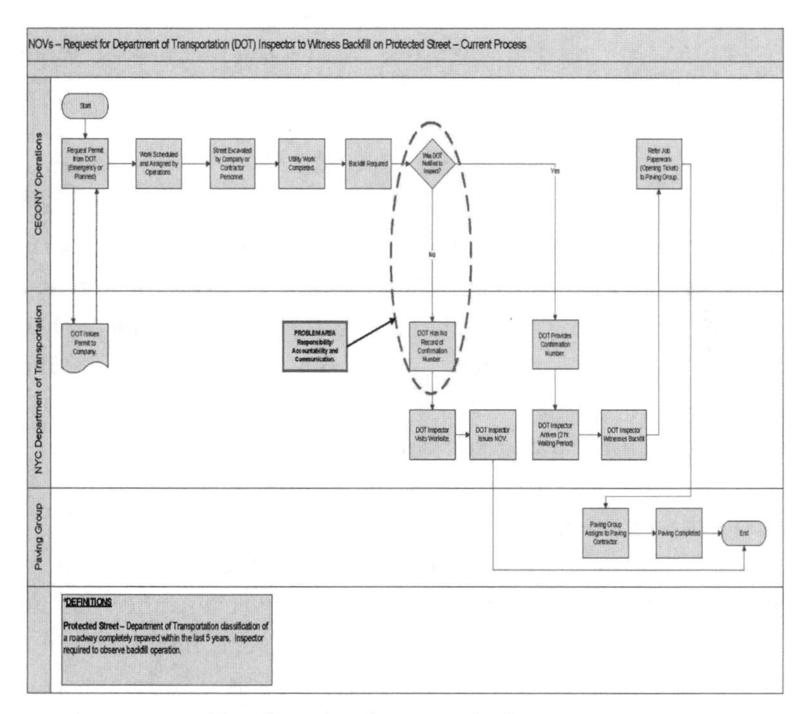

FIGURE 10.11. Key elements of current city inspector request process.

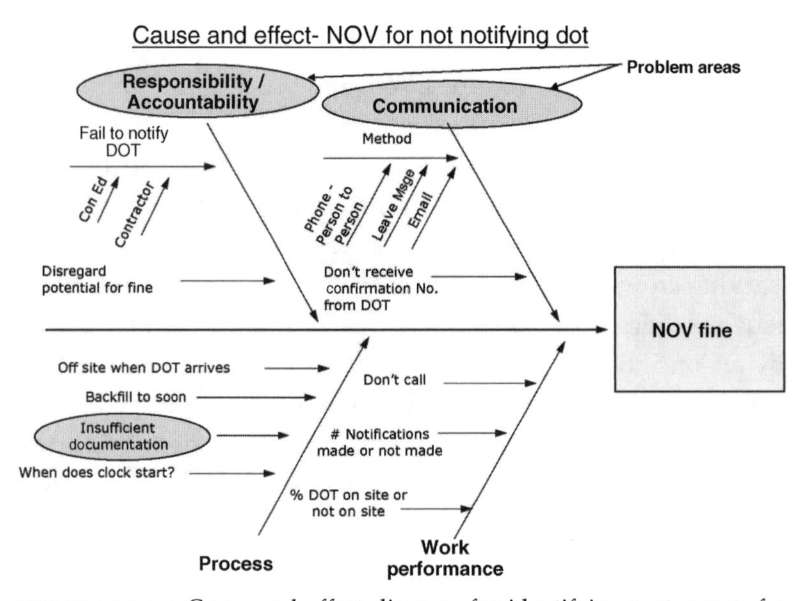

FIGURE 10.12. Cause and effect diagram for identifying root causes for failing to request city inspectors.

contractor. In the third case, the work is done solely by a Con Edison crew. Regardless of who does it, however, it is unclear who is responsible to alert the city to request an inspector. Should it be a crew member or someone back in the office? If a combination of different crews performed the job, which crew is responsible for making the contact? If no one is responsible, then no one is held accountable. This glitch led immediately to the next root cause.

The failure to alert the city, followed by the resulting fine, is often caused by a failure to communicate. The utility has no consistent process for alerting the city to send inspectors in a timely fashion. What at first blush appears to be a simple issue is complicated by the use of contractors and by the possibility of having these communications originate from a string of different Con Edison departments.

Crews did not always inform operating areas about the status of a job, especially during backfilling. Additionally, operating areas did not always keep abreast of what field crews were doing, leading to having backfilling completed without alerting the city. Months or even years later, the DOT would review paperwork and often discover that inspectors had not witnessed many backfills. After visiting the site and finding the work had been completed, DOT issued a violation. The team also discovered that some Con Edison crews or contractors had actually notified DOT, but an inspector had never appeared. Unfortunately, those who placed the call rarely kept accurate records. Consequently, Con Edison could not provide supporting documentation to waive the fine.

No single form was used by all Con Edison organizations to record a request for an inspector. Records were spotty and inconsistent. An employee might record the date but not the time a call was made, neglect to indicate who made the call, or fail to record whether a message was left with the city or if anyone had spoken to city employees to obtain a confirmation number. All of these pieces of information were needed to defend the utility from receiving improper fines.

Given the inconsistencies discovered, the team proposed both short-term and long-term solutions. In the short term, a standard request form collecting pertinent data with confirmation numbers should be used uniformly throughout Con Edison. The group also suggested that responsibility for contacting the city be with a Con Edison staff member in each operating area. By implementing these recommendations, all employees would know who is responsible for requesting an inspector and who was accountable should a violation be issued.

For the longer term, the team proposed an automated solution. The transportation department had originally set up one number for calls covering everything from emergency permits to emergency authorization numbers and inspectors.

After several utilities complained about difficulties in leaving messages or getting someone on the phone, the DOT established separate numbers for each borough. While this option made it slightly easier to call in requests, the system was still incapable of handling the volume.

That's why the team recommended that Con Edison's Information Technology department collaborate with the city to create an electronic system linking Con Edison to the DOT. The new system would be modeled after a current system designed for requesting emergency authorization numbers. Once the system is in place, Con Edison operating areas would log on, complete an appropriate form, and send it off electronically to the department. DOT would automatically record the request, alert an inspector, and return a confirmation number, altogether eliminating phone calls, dispensing with reaching a "live" person on the phone, and ending the need for special forms.

Team members then worked on creating a standardized request form for the various operating areas of the company. During ensuing discussions between Con Edison and DOT, the city concluded that an automated system would be helpful for all companies working in New York City, not just Con Edison. In the end, they introduced a new electronic system for requesting inspectors and receiving confirmation numbers.

As of January 2009, online submissions are the only way to request and confirm a DOT inspector for backfilling excavations. Key Con Edison staff—including several who were members of the project team—are currently demonstrating the use of the new system and disseminating information about it to the various areas of the company that engage in street construction and excavation. In the meantime, it is projected that the utility will save approximately $150,000 annually.

DEWATERING CHALLENGES IN CONSTRUCTION PROJECTS

Municipalities, such as New York City, have many proposed public improvement projects scheduled each year. During initial review, it is usually discovered that there are existing structures as well as infrastructures in the area blocking implementation. To clear a path for new projects, the city alerts the relevant agencies and private companies of an impending project and gives them a chance to remove what may be in the way. These are known as "interference" projects.

While performing interference work for the City of New York, Con Edison often suffers damages due to an unforeseen groundwater conditions at the site. The city may also allege that contamination has leached from the existing Con Edison structures. The utility then incurs added costs and penalties for time delays, excessive wastewater-handling costs, other environmental cleanup costs, and potential environmental violations from city, state, and federal agencies. A virtual team of engineering and Environmental Health and Safety staff collaborated on analyzing process planning and suggestions for resolving environmental conditions in public improvement construction projects.

During a root cause analysis, the team used a process flowchart, manager interviews, industry-accepted service standards and measures, and a House of Quality method to map the company's current internal process for performing interference work. The team focused on identifying and resolving internal and external customer needs. During an evaluation of the city's prior notice schedule, the team discovered that several key Con Edison departments were not receiving notices of proposed interference projects until the projects were well under way. The team also reviewed project cycle times, the percent of major projects that receive due diligence and hydrogeology testing, the penalties for project delays, and fines associated with environmental or permit violations. They also found that company environmental procedures did not require detailed reviews of these projects (Figure 10.13).

The team uncovered several areas in which Con Edison's interference project process required improvement. For example, the company's cross-functional distribution list needed to be updated to include all interested parties. The team suggested that environmental due diligence tools require standardization, including checklists of environmental procedures to cover hydrogeological testing (Figure 10.14). The team also proposed introducing steps to ensure the city's final design drawings, environmental impact statement, and Con Edison's current field observations or test pits are the basis for all decisions concerning the proposed project. They also suggested that the utility's Engineering and Construction departments be given

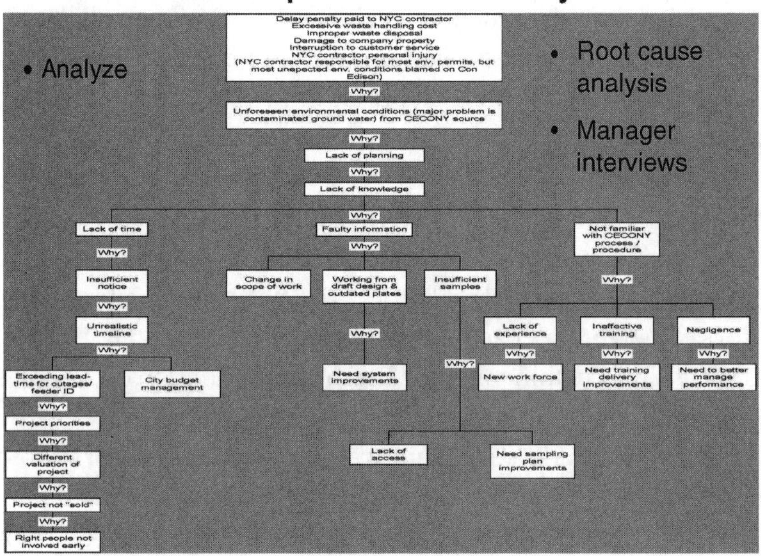

FIGURE 10.13. Analysis of current public improvement project process.

Better Project Planning

Perform Due Diligence (CEP 11.03/11.04) for environmental conditions.

CEP 11.03 excerpt

Section 28, Part I: Water

- Will dewatering operations be required either during construction or when the facility is completed?
  Yes ☐ **No** ☐ Examples of dewatering operations that may be needed:
  - Construction involving excavations or trenches in which rainwater or groundwater can accumulate,
  - Oil-water separators,
  - Storm water management,
  - Hydro testing new pipelines.
- Will the project disturb one (1) acre or more of land? **Yes** ☐ **No** ☐
- Are dry wells present on site? **Yes** ☐ **No** ☐

Check the box if you answered Yes to any question. ☐

Complete Entire Section 28, Part II: Water

CEP 11.04 excerpt (New due diligence data)

Topic	**Check (✓)**	**If yes, describe the actions taken to address this concern.**
28. Water	**Yes** ☐ **No** ☐	
Historical		
MGP Site?	**Yes** ☐ **No** ☐	
Appendix B?	**Yes** ☐ **No** ☐	
Incident Reports at site?	**Yes** ☐ **No** ☐	
Utility Plates reviewed?	**Yes** ☐ **No** ☐	
ChemLab Analyticals?	**Yes** ☐ **No** ☐	
EDR Search performed?	**Yes** ☐ **No** ☐	
Hydrogeological		
Depth of water table	________**FT.**	
Tidal Charts reviewed?	**Yes** ☐ **No** ☐	
Presence of underground streams – Viele Map?	**Yes** ☐ **No** ☐	
Is site in flood zone?	**Yes** ☐ **No** ☐	
Is site a wetland - Wetland Maps?	**Yes** ☐ **No** ☐	

FIGURE 10.14. Proposed change to corporate environmental procedure.

pre-excavation environmental sampling and analysis as well as perform pre-planning for the best options for treatment and disposal of contaminated groundwater. This last improvement builds on the company's considerable expertise in wastewater treatment gained from underground flush operations, steam stations, and other remediation projects. It would also provide various cost–time–compliance options for the treatment and disposal of contaminated groundwater.

Today, the team not only continues to work on solutions for interference projects, but also the scope of its work has expanded now to include all construction projects. They have completed feasibility and cost analyses on various options for treating groundwater (Figure 10.15) and will begin working with the process owners and Corporate Environmental Health & Safety personnel to implement their recommendations. A pilot test is being planned at a remediation site together with the city's Department of Environmental Protection. The team also plans to explore options for general permits and pre-approvals based on likely scenarios with various local agencies.

Portable Water Treatment Unit

- Cost Comparison

Treatment Method		Cost for Rental	Cost per Gallon	Transportation /Mobilization Cost	Total Cost (assuming one week rental)
Portable Water Treatment Unit	Siemens	$10K for two weeks	$0.25 based on 40,000 gal.	Unknown	$10K + operator
	Veolia	$15,000 for one week with cation exchange	$0.38 based on 40,000 gal.	Unknown	$15K
	TIGGS	$8k - 10K per month	$0.25 based on 40,000 gal.	Unknown	$10K + operator
Transport by Con Ed and dispose to Astoria		None	$0.30 per gallon	Direct charge back to account	$12K + $90/hr.
Transport by vendor and dispose at outside disposal facility		None	$0.48 per gallon (Allstate)	$750 per tanker + cost of frac tanks	$22K (based on 4 loads)

FIGURE 10.15. Initial cost comparison of portable water treatment systems.

Gas Mapping Process Improvement

One of Con Edison's principal businesses is the delivery of natural gas to approximately 1.2 million customers in New York City and Westchester, Orange, and Rockland Counties. In this 660 square mile area, there are 4240 miles of gas main. Naturally, Con Edison crews and contractors constantly perform routine maintenance and emergency work on its gas infrastructure. During the work, crews that install new facilities or discover incorrectly mapped structures note the changes and route them to Gas Engineering and the Maps & Records department in order to keep records updated. Routine updated records decrease future damage to the company's gas infrastructure by vendors, other utilities, and private businesses. But because of the frenetic pace crews often experience, it takes considerably more time to update records than is often available, leading at times, unfortunately, to infrastructure damage. To appreciate the problems, virtual team members from Gas Construction and Gas Engineering collaborated to reveal the process for mapping gas facilities (Figure 10.16).

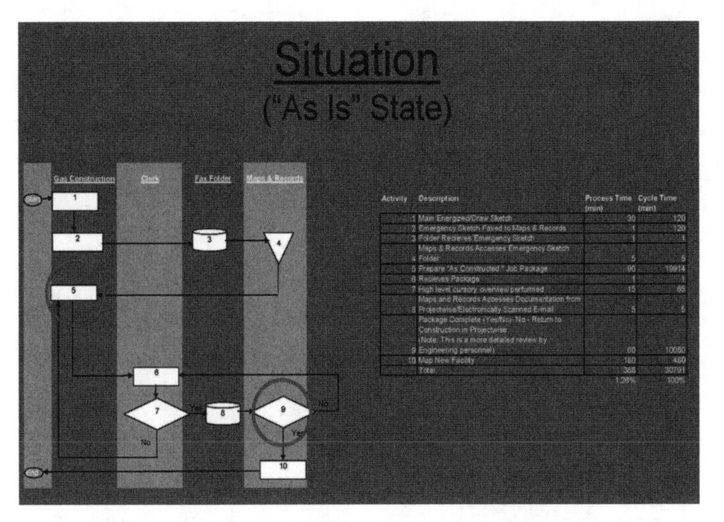

FIGURE 10.16. Flowchart for current gas mapping process.

The team found that emergency sketches require 24 hours to produce and an additional two weeks to send packages to and from Maps & Records. Mapping gas facilities often required another two weeks. The Maps & Records department manually counted the packages received, the number that arrived ahead of schedule, as well as those that were incomplete or inaccurate.

To learn where the bottlenecks occurred, the team mapped the process into a cause and effect diagram (Figure 10.17). It found four broad areas that contributed to the problems encountered. The team noted that there is no uniform process used by all operating areas for submitting as-constructed drawings.

No uniform and easily tracked process existed for submitting emergency drawings. Newly hired supervisors did not receive adequate training to understand the necessary documents required for all field records. They also found that there was no adequate or timely way to notify operating areas that packages contained incorrect or deficient information.

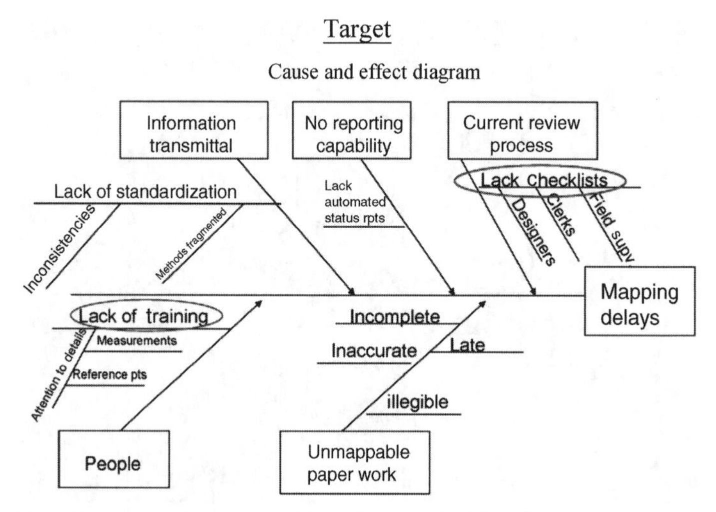

FIGURE 10.17. Cause and effect diagram to identify current process.

Once these problems were identified, the team designed a revised workflow diagram incorporating their recommendations (Figure 10.18).

They suggested a uniform repository process for Gas Operations departments in order to decrease errors and lost

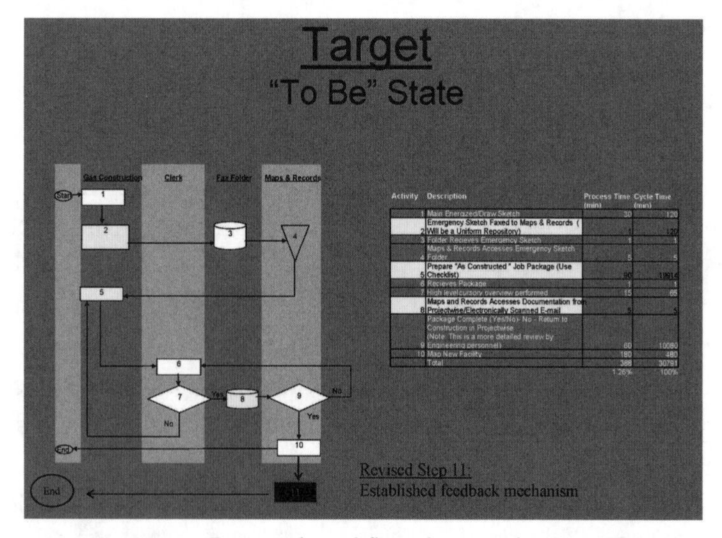

Activity	Description	Process Time (min)	Cycle Time (min)
1	Main Energized/Draw Sketch	30	120
2	Emergency Sketch Faxed to Maps & Records (Will be a Uniform Repository)	1	120
3	Folder Recieves Emergency Sketch	1	1
4	Maps & Records Accesses Emergency Sketch Folder	5	5
5	Prepare "As Constructed" Job Package (Use Checklist)	90	19914
6	Recieves Package	1	1
7	High level/cursory overview performed	15	65
8	Maps and Records Accesses Documentation from Projectwise/Electronically Scanned E-mail	5	5
9	Package Complete (Yes/No)- No - Return to Construction in Projectwise (Note: This is a more detailed review by Engineering personnel)	60	10080
10	Map New Facility	180	480
	Total	388	30791
		1.26%	100%

FIGURE 10.18. Proposed workflow diagram for gas mapping.

paperwork and allow Maps & Records to easily access information. The team proposed a checklist, accompanied by templates, to assist administrative teams in standardizing job packets, decreasing cycle time, and enabling reviewers to pinpoint delinquent packets immediately. They also proposed a feedback mechanism as a form of internal quality assurance to ensure that mapping is performed and to reduce the amount of lost paperwork. The proposal also included a training plan for field workers in order to educate them about information required for mapping facilities, thus increasing the quality of work, decreasing errors, and reducing the overall cycle time.

The virtual team, together with other staff from Maps & Records and training personnel, is now in an implementation phase. Currently, training for new supervisors and clerks includes additional work on gas mapping requirements. Gas Operations now works under a uniform repository system for all as-constructed drawings, which produces a side benefit of an audit trail for all documents.

TIPS FOR SUCCESS WITH CORPORATE VIRTUAL TEAMING

The virtual team experience can be successful for everyone if a few guidelines are followed:

Tip 1: Pre-select actual business cases for class analysis. Solicit projects and business processes from directors, managers, field workers, and other personnel throughout the company. Actual users are often the ones who suspect that a current business system could be improved.

Tip 2: Select a cross-functional team with some familiarity with the business case. Ideally, a team should be comprised of five to six members from various job functions and departments. If the business process interacts with several departments within the company, staff from each should be on the team. Remember that an individual does not have to be intimately familiar with the process to be a valuable team member. Those who understand the terminology, but rarely interfaces with the process itself, are not hindered by preconceptions or the "normal way we do business."

Tip 3: Explain the purpose of virtual teaming and the analytical tools that will be used. Team members need to understand and "buy in" why virtual teaming is important in today's business climate. They also need to feel that help is available to them whenever it is needed.

Tip 4: Help those who have never experienced virtual teaming or learning. Some may be uncomfortable or inexperienced with virtual learning. Taking extra time with these individuals often results in a fuller team experience for everyone.

Tip 5: Interact and communicate. Communication is a key element in virtual teaming. All team members must be engaged in the project and "talk" with each other regularly. Physical meetings between team members can also be useful, but the use of various modes of communication effectively and efficiently should be encouraged.

Tip 6: Divide work among all team members. Engage everyone in the analysis. Make everyone responsible for a deliverable. Make sure the team understands that leaders and loafers exist in the virtual world, too.

Tip 7: Each team member's supervisor or manager must buy into the project. The virtual team world also has responsibilities, deadlines, and deliverables. Each team member will need to devote time and resources to the project just like any other. Supervisors and managers need to understand that an employee's time spent in virtual teaming can yield significant results for the company as well as expand an employee's knowledge and confidence levels over the course of the project.

Collaboration between Stevens and Con Edison's senior management has given the company ownership in the course's success. Virtual team recommendations continue to be implemented and participants are nominating others for the next set of courses.

Con Edison fosters a highly analytical culture and relies heavily on data and analyses in making business decisions. A course that explores techniques to arrive at decision-based models and incorporates them into actual work problems enhances the utility's day-to-day operations. Some virtual team members continue to exploit the analytical tools they learned to use in the course to drive process change in their organizations. Examples include reducing time to solve complicated telecommunications problems from four to two days, analyzing feeder processing time to reduce customer outage time, reducing paper-intensive processes by half and achieving a 30% decrease in overall costs, and undertaking additional process reviews for other city violations.

Overall, virtual team participants have employed analytical techniques from the course to produce measurable and significant recommendations and changes at Con Edison. The success of e-learning and the virtual classroom have made it easier for larger and varied groups of employees to benefit from their participation, without the added travel expense and work stoppage of a conventional on-ground classroom experience.

Con Edison is a regulated utility that provides electric service in New York City and most of Westchester County. The company provides natural gas service in the New York City area and Westchester. Con Edison also owns and operates the world's largest district steam system, providing steam service in most of Manhattan. It has approximately 15,000 employees and 14,000 retirees. To help train and educate employees, the company has created The Learning Center, a corporate university and training facility located in the Long Island City section of Queens, NY. The Learning Center offers approximately 800 courses in both skills training and leadership development.

CHAPTER

11

VIRTUAL ENGINEERING TEAMS[1]

DOUG VOGEL, MICHIEL VAN GENUCHTEN, CAROL SAUNDERS, AND A.-F. RUTKOWSKI

With the introduction of new modes of communication over the last few decades, vast changes have taken place in engineering. Large numbers of projects are now globally distributed, with significant activities in the Far East, especially in India and China, while headquarters operations remain in the West. Bridging the divide is now crucial to managing projects successfully (Hofstede, 1980).

Today, engineering projects rely heavily on professional communication, exploiting a portfolio of interactive tools throughout the design phase (van Luxemburg et al., 2006; Pearlson and Saunders, 2006). Software now plays an increasingly central role as face-to-face communication becomes less frequent. A wide range of communication technologies—such as voice over IP, e-mail, and chat—now bridge distances and time zones.

While e-mail and audio- and videoconferencing are essential communication tools across time zones, disciplines, and

[1]Adapted from Rutkowski, A. F., Vogel, D., van Genuchten, M., and Saunders, C. (2008) Communication in virtual teams: 10 years of experience in education. *IEEE Journal on Professional Communications*, 51(3), 302–312.

Virtual Teamwork: Mastering the Art and Practice of Online Learning and Corporate Collaboration. Edited by Robert Ubell

national boundaries, these rich, technologically enhanced experiences are largely absent from traditional engineering classrooms. In a typical 100-hour course, it is unlikely that students will be exposed not only to the core curriculum, but also to real-world global projects. Even as international programs at colleges and universities expand, the opportunity to meet and collaborate with students from different cultures continues to be quite limited. Achieving the principal goal of an engineering education—building a solid core of knowledge—combined with student collaboration across professional cultures has largely been ignored in traditional classrooms.

How can real-life engineering management be taught? The ancient Chinese philosopher Lao Tzu remarked, "If you tell me, I will listen; if you show me, I will see; but if you let me experience, I will learn." While there are many ways to provide students with real-life engineering experiences (Gillet et al., 2005; Vallim et al., 2006), one option—the approach chosen by most schools—is to scale down an actual project to a size that fits within a typical 100-hour workload. The second is to combine the 100 hours with 100 or more students to create a 10,000-hour project, establishing an environment in which students have a greater likelihood of experiencing real-life engineering.

Virtual teams, supported by electronic communication, have been introduced for a decade to increase student awareness of working in a global context (Jarvenpaa and Leidner, 1999; Piccoli et al., 2001; Vogel et al., 2001; Hornik et al., (in press); Vaverek and Saunders (1993–1994)). Several reviews report on a large number of such team-based courses (Martins et al., 2004; Powell et al., 2004). While virtual teaming may extend student access to different cultures, they do not typically participate in large, complex projects. Much of the research reports on teams that meet on an average just four or five weeks (Powell et al., 2004), working on short-term tasks only (Martins et al., 2004).

HONG KONG NETHERLANDS PROJECT

The Hong Kong Netherlands project (HKNet) is an example of integrated learning among multiple international institutions, providing students with the reality of engineering management, coupled with professional communication in college and university settings. Courses at participating universities range from software management to information systems development, with a software project chosen for large-scale student collaboration. Students engage in real-life planning, problem solving, and the selection of a portfolio of technologies to achieve high-level performance.

The principal aim is to provide students with the opportunity to work on a project at the level of complexity that they are likely to find in jobs once they graduate. It exposes students to different cultures with guidance in resolving conflicts that may arise from cultural diversity. The project gives students the experience of working remotely on global virtual teams, allowing them to select and implement a range of communication media to implement the project successfully.

VIRTUAL ENGINEERING CLASSROOM

In the first years, students study core constructs and theory individually. They also engage in continuous interaction with students from different professional and national cultures. The goal is to deliver a high-quality product in the form of an electronic book. Over a 10-year period, more than 1000 students participated (Vogel et al., 2001; Rutkowski et al., 2002, 2007; van Genuchten et al., 2005).

About half the participants are part-time business students and the other half full-time university engineering students. While the project started with just two universities—Eindhoven and Hong Kong—over the years, it attracted a number

Group 2: Risk Management

FIGURE 11.1. Morphed picture of an HKNet team in 1998.

of others, expanding to include the universities of Tilburg (The Netherlands), Beijing (China), and Orlando (USA). At any particular moment, the number of students ranged from 65 to 180, depending on enrollment at various sites.

Typically, each HKNet team consisted of 8–10 students in two to four locations. Figure 11.1 shows an example of a 1998 team. Students never met face-to-face. Each team was assigned a software-related topic—say, "the impact of software defects," "television on mobile phones," or "open source and software patents." Over eight weeks, students viewed topics from their different geographical perspectives, engaging in a structured process with divergent and convergent activities. Teams were asked to build web sites covering their topics to illuminate European and Asian perspectives. Web sites, drawn from all groups, were then integrated into an electronic book, an exercise that not only demonstrated their web site-building skills, but also gave them the opportunity to become experts in their specialized content area.

The multiyear process, illustrated in Figure 11.2, provides students with a repository of work accomplished in previous years for them to draw upon in the current year. Students are

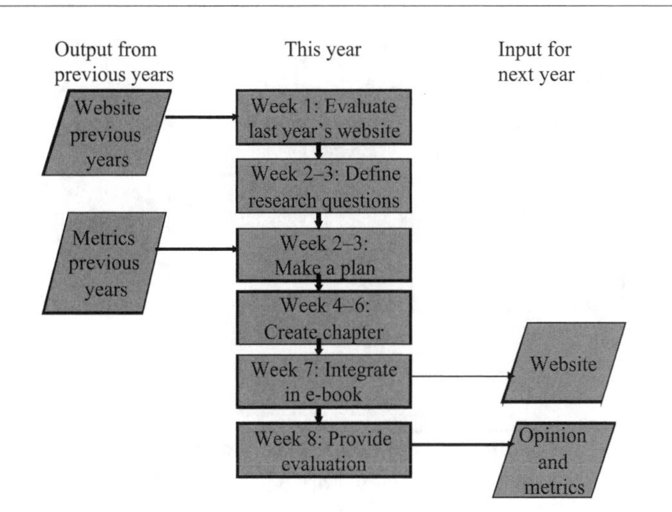

FIGURE 11.2. The learning process.

required to define a number of research questions, devise a plan, and collect study materials. At the end of six weeks, they are required to deliver their team's chapter covering the research topic they selected. By week seven, they are expected to integrate it into an e-book. Students then critique each other's e-book chapters, comment on team processes, and evaluate the contributions of each team member. A wrap-up report concludes the project.

To demonstrate project outcomes, results are presented in a web site that forms an electronic book of the software industry. Found at http://www.bohknet.com, a screenshot of the homepage is shown as Figure 11.3. BOHKNet stands for Beijing, Orlando, Hong Kong, and the Netherlands. The site also offers more information about the project, including videos of lectures.

COMMUNICATION

Videoconferencing, e-mail, Second Life, and Blackboard, a learning-management system that supports both chat and

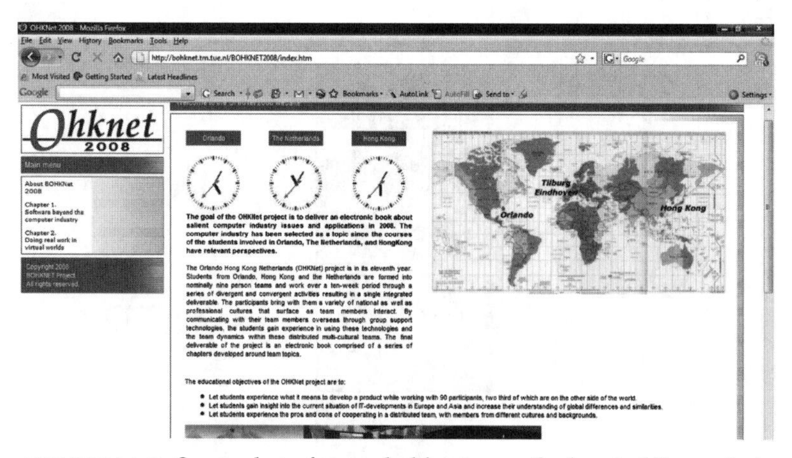

FIGURE 11.3. Screenshot of www.bohknet.com, the heart of the project.

forums, were employed as communication vehicles. Videoconferencing is introduced at the start of the project and halfway through again, allowing students to track their progress and giving them a chance to resolve disputes. A final videoconference is held toward the end to celebrate results.

Working from remote locations in virtual teams is a fact of life in industry today. Virtual teams combine the classical challenges of face-to-face groups as well as less traditional obstacles of computer-based interaction in complex environments. Research has shown that for complex tasks, such as software design and inspection, virtual teams can outperform conventional ones in terms of quality and creativity (Ocker et al., 1995; van Genuchten et al., 2001). Research has also demonstrated that web-based applications can efficiently support distributed teamwork when critical aspects of social interaction are taken into account (Hollingshead and McGrath, 1995). National culture, trust, temporal coordination, leadership, networking, social loafing, and group history are some of the factors that can undermine or help distributed teams succeed.

COMMUNICATION PLANNING

Project planning is typically part of most engineering courses. To help students plan, HKNet asks them to provide an estimate of the number of screens proposed for their web site, a breakdown of the work to be performed, including who does what, and how much effort is involved. They are also asked to predict the top five risks, estimating likelihood and potential impact.

In an introductory lecture, an instructor illustrates how to plan, providing metrics generated by earlier HKNet projects, such as the number of web site screens and effort required. The recorded lecture is available online for students at various locations, giving members of all teams the same background information.

HKNet student plans tend to be quite different from exercises performed in conventional classes; the consequences of devising a plan for a fictitious project differ markedly from one on which students will be working for the next weeks. For many, it is the first time they have experienced what it means to negotiate with team members to create a working plan. When student teams draw up a well-conceived plan, it increases the likelihood of success and reduces conflict.

Projects in industry often require participation from multiple locations and cultures. Today, global redistribution of engineering work has led to the participation of significant numbers of Indian and Chinese engineers. While some courses offer curriculum on cultural differences and how to handle the challenges of working together (Koen, 2005), other programs may ask students to role-play to represent their own or a remote culture. Obviously, role-playing is far less realistic than working with your cultural counterparts over an eight-week period. Working together under the guidance of faculty encourages students to share their reactions and to work through problems. Without guidance, teams may fail to recognize the complexities of cultural difference and deteriorate into conflict (Anawati and Craig, 2006).

Communication Technologies

While the present student generation has most likely spent more time behind a computer than a television screen, and while messaging and chat may be routine for them, working on delivering an engineering result is quite different. At HKNet, students use Blackboard as their central learning site. They are free to use other tools, such as chat, Internet telephony, and videoconferencing (Saunders et al., 2006). It is often the first time that many have had to select computer-based tools and live with the consequences in a real project.

Students learn about the importance of task–technology fit. Following the well-known DeSanctis and Poole (1994) model, students come to realize that their videoconference sessions will be unproductive without sufficient preparation. They discover that putting aside time to prepare discussion documents is the key to success.

Assignments and Grades

In the traditional classroom, instructors can closely follow the progress of their students. In virtual classes, assignments and grading require special attention. To minimize social loafing, recognized as the dark side of virtual collaboration (Jarvenpaa and Leidner, 1999), at HKNet, students are given two assignments based on the notion of social facilitation and social loafing (Harkins and Szymanski, 1987), as well as on Lewin's (1948) field theory. In week seven, the team is presented with a task designed to be both challenging and attractive (Harkins and Petty, 1982), requiring team members to interact with one another to construct knowledge collaboratively. On completion, all team members receive the same grade—determined by a pool of five independent instructors who evaluate the quality of the team's web portal and rank it in comparison with the results of other virtual teams. The assignment is complex, and requires pooled contributions and a high sense of coordination. It also generates intergroup competition among the teams, with a reward to the best virtual team as an incentive.

In the second assignment, each student prepares a critique of chapters in the e-book, evaluating the process that was followed by their own virtual team, and reporting on the contributions of each member. The second assignment is more traditional, with individual student work clearly identified and rated independently. It serves as a back-up, in case of major team project conflict or failure. Students are given a chance to offer their own reflections on what constitutes a successful solution. On the whole, students are eager to receive recognition for their individual contributions. This assignment gives those who made special contributions to the team's work to be rewarded; it also penalizes those who failed to participate collaboratively.

PORTFOLIO OF TECHNOLOGIES

It is especially important to provide a portfolio of applications from which students can choose those that best fit their needs. It helps identify technologies that offer benefits to team members across time and space. It turns out that teams frequently have different technology preferences, occasionally the result of subgroup culture (Sivunen and Valo, 2006). For example, US students are usually eager to use social networking tools, such as MySpace and Facebook, to help members get to know one another. Multinational teams, on the other hand, often turn to Blackboard as the site of preference to introduce themselves.

Students from the Netherlands and Hong Kong rarely post their personal information on social networking pages.

In another example, a subgroup, wishing to reduce uncertainty or concerned about loss of face in synchronous interaction, might prefer an asynchronous solution. Another subgroup, hoping to move quickly to resolve a dispute, may select a synchronous option. Teams tend to converge rather quickly on a solution that broadly meets subgroup needs, tempering individual preferences to benefit the team as a whole. Teams tend to establish their own way of working together, often using communication tools fairly creatively.

COOPERATIVE TECHNOLOGIES

Begun in 1998, HKNet at first used the groupware tool, GroupSystems, running on a first-generation thin-client technology. It was built originally for face-to-face meetings. Figure 11.4 shows students at work in the Boardroom using GroupSystems.

During the first few years of the project, students used e-mail and instant messaging, in addition to groupware. A few

FIGURE 11.4. HKNet 1998 boardroom.

Forum

Display Order	Forum	Total Posts	Unread Posts	Total Participants
1	Team Member Introductions Please use this forum to tell something about yourself and learn more about your team members. Hint: Start a new discussion thread for each team member to keep things better organized.	4	2	4
2	Team Planning Use this forum to let team members know when you will be working on the project. Your team member can see when you will be working on the project for e.g. a simultaneous session. If you are absent for a long period e.g. business trip, fill in when so that your team members know.	1	1	1
3	Project Discussion Use this forum to discuss issues regarding the total management of your project. Any remarks, suggestions and questions about issues concerning your whole team like the progress, planning, approach and division of work can be discussed here. Hint: start a new thread for each new topic to help manage the communication space.	11	4	5
4	Potential Research Questions What are some research questions associated with your topic?	19	0	7
5	Agreed Upone Research Question 1 What are the issues that need to be addressed in this question and who will be responsible?	6	0	3
6	Agreed Upon Research Question 2 What are the issues that need to be addressed in this question and who will be responsible?	7	1	2
7	Agreed Upon Research Question 3 What are the issues that need to be addressed in this question and who will be responsible?	24	3	3
8	Deliverables Use this forum to begin to formulate your deliverables.	47	25	11
9	Photos Photos from your team interactions	2	2	1

FIGURE 11.5. Screenshot of a 2008 forum.

years later, HKNet4 introduced Blackboard, an environment that offered students a place to store their results and delivered synchronous and asynchronous communication. It also provided faculty with tools to track student participation. A screenshot of a forum discussion board is given in Figure 11.5. Recently, students have been migrating to free communication and cooperation tools, such as MSN, Skype, and Google Docs.

Second Life was introduced more recently. An Alpine Executive Centre http://slurl.com/secondlife/MeetingSupport/116/54/21/?i&title=Alpine%20Executive%20Center, depicting HKNet activities in Second Life, is shown in Figure 11.6. On the left is a fantasy aerial view of the Alpine Executive Centre. The middle illustration shows a more informal simulation of students and faculty. An island in Second Life, the Alpine Executive Centre is a space to hold virtual meetings. Tucked away in an alpine ski

FIGURE 11.6. HKNet in Second Life.

village, surrounded by snow-covered mountains, lies an advanced meeting facility where real-world activities take place in a virtual environment. An amphitheatre is accessed by a train that travels deep inside the mountain complex or you may walk along a frozen ice-skating pond, or travel by teleport from a visitor landing area. An auditorium supports large groups in plenum sessions for presentations and moderated discussions. It is shown at the right. A host of additional facilities exist to support groups. For example, groups can meet at 1 of the 10 gathering spots around the village, including mountain huts with interactive screens and scenic lookouts. Figure 11.7 shows a Dutch and a HK student playing the piano in their virtual meeting place.

To a large extent, the choice of technology is left to the students. They must not only select the tools, but also live with the consequences. While eventually we expect that virtual worlds will introduce significant communication benefits, because the application is in a pioneering phase, it is unclear how well it can be integrated with other applications on the Internet. What's more, because virtual words are used synchronously, students in far-off time zones cannot join in easily.

Unquestionably, HKNet students benefited from taking this virtual teaming course, especially when they enter the

FIGURE 11.7. Hong Kong and Netherlands students playing the piano in a virtual world.

world of work after graduation. Nowadays, job applicants are commonly asked whether they have had any virtual-team experience. Some companies hold job interviews by phone or videoconference to demonstrate an applicant's skill in communicating virtually.

Ten years of experience with HKNet shows that it is possible to bring real-life engineering into a university setting. It demonstrates that it is possible to give students experience working and communicating as part of a multinational virtual team that produces a real product. The web site, www.bohknet.com, archives the engineering education materials.

REFERENCES

Anawati, D. and Craig, A. (2006) Behavioral adaptation within cross-cultural virtual teams. *IEEE Transactions on Professional Communication*, 49(1), 44–56.

DeSanctis, G. and Poole, M. S. (1994) Capturing the complexity in advanced technology use: adaptive structuration theory. *Organization Science*, 5(2), 121–147.

Gillet, D., Ngoc, A., and Rekik, Y. 2005 Collaborative web-based experimentation in flexible engineering education. *IEEE Transactions on Education*, 48(November), 696–704.

Harkins, S. and Szymanski, K. (1987) Social facilitation and social loafing: new wine in old bottles. *Review of Personality and Social Psychology*, 9, 167–188.

Harkins, S. G. and Petty, R. E. (1982) Effects of task difficulty and task uniqueness on social loafing. *Journal of Personality and Social Psychology*, 43, 1214–1229.

Hofstede, G. (1980) *Culture's Consequences: International Differences in Work-related Values*. Newbury Park, CA: Sage.

Hollingshead, A. B. and McGrath, J. E. (1995) Computer-assisted groups: a critical review of the empirial research. In: Guzzo R.A. and Salas E., (Eds.) *Team Effectiveness and Decision-Making in Organizations*. San Francisco CA: Jossey-Bass, pp. 269–304.

Hornik, S., Saunders, C., Li, Y., Moskal, P., and Dzubian, C. (2008) The impact of paradigm development and course level on performance in technology-mediated learning environments. *Informing Science*, 11, 35–58.

Jarvenpaa, S. L. and Leidner, D. E. (1999) Communication and trust in global virtual teams. *Organization Science*, 10(6), 791–815.

Koen, B. (2005) Creating a sense of presence in a web-based PSI course: the search for Mark Hopkins' log in a digital world. *IEEE Transactions on Education*, 48(November), 599–604.

Lewin, K. (1948) *Resolving Social Conflicts*. New York: Harper.

Martins, L. L., Gilson, L. L., and Maynard, M. T. (2004) Virtual teams: what do we know and where do we go from here? *Journal of Management*, 30(6), 805–835.

Ocker, R., Hiltz, S. R., Turoff, M., and Fjermestad, J. (1995) The effects of distributed group support and process structuring on requirements development teams: results on creativity and quality. *Journal of Management Information Systems*, 12(3), 127–153.

Pearlson, K. E. and Saunders, C. S. (2006) *Managing and Using Information Systems: A Strategic Approach*. New York: John Wiley and Sons.

Piccoli, G., Ahmad, R., and Ives, B. (2001) Web-based virtual learning environments: a research framework and a preliminary assessment of effectiveness in basic IT skills training. *MIS Quarterly*, 25(4), 401–426.

Powell, A., Piccoli, G., and Ives, B. (2004) Virtual teams: a review of current literature and directions for future research. *The Database for Advances in Information Systems*, 35(1), 6–16.

Rutkowski, A.-F., Saunders, C., Vogel, D., and van Genuchten, M. (2007) Is it already 4 a.m. in your time zone? Focus immersion and temporal dissociation in virtual teams. *Small Group Research*, 38(1), 98–129.

Rutkowski, A.-F., Vogel, D., van Genuchten, M., Bemelmans, T., and Favier, M. (2002) E-collaboration: the reality of virtuality. *IEEE Transactions on Professional Communication*, 45(4), 219–230.

Saunders, C. S., Sivunen, A., and Valo, M. (2006) Team leaders' technology choice in virtual teams. *IEEE Transactions on Professional Communication*, 49(1), 57–68.

Sivunen, A. and Valo, M. (2006) Team leaders' technology choice in virtual teams. *IEEE Transactions on Professional Communication*, 49(1), 57–68.

Vallim, M. B. R., Farines, J. M., and Cury, J. E. R. (2006) Practicing engineering in a freshman introductory course. *IEEE Transactions on Education*, 49(February), 74–79.

van Genuchten, M., van Dijk, C., Scholten, H., and Vogel, D. (2001) Using group support systems for software inspections. *IEEE Software*, (May–June), 18(3), 60–65.

van Genuchten, M., Vogel, D., Rutkowski, A.-F., and Saunders, C. S. (2005) HKNET: instilling realism into the study of emerging trends. *Communications of AIS*, 15, 357–370.

van Luxemburg, A. P. D., Ulijn, J. M., and Amare, N. (2006) The contribution of electronic communication media to the design process: communicative and cultural implications. *IEEE Transactions on Professional Communication*, 49(4), 250–264.

Vaverek, K. and Saunders, C. (1993–1994) Computerspeak: message content and perceived appropriateness in an educational setting. *Journal of Educational Technology Systems*, 22(2), 123–139.

Vogel, D., van Genuchten, M., Lou, D., Verveen, S., Eekhout, M., and Adams, T. (2001) Exploratory research on the role of national and professional cultures in a distributed learning project. *IEEE Transactions on Professional Communication*, 44(2), 114–125.

INDEX

Virtual Teamwork: Mastering the Art and Practice of Online Learning and Corporate Collaboration. Edited by Robert Ubell